Application of Computational Electromagnetics Techniques and Artificial Intelligence in the Engineering

Application of Computational Electromagnetics Techniques and Artificial Intelligence in the Engineering

Editors

Le Xu
Xiaowei Shi
Rui Li
Xiaojie Chen
Xiaoqun Chen

MDPI

Basel • Beijing • Wuhan • Barcelona • Belgrade • Novi Sad • Cluj • Manchester

Editors

Le Xu
Xidian University
Xi'an
China

Xiaowei Shi
Xidian University
Xi'an
China

Rui Li
Xidian University
Xi'an
China

Xiaojie Chen
CAEP Software Center for High Performance Numerical Simulation
Beijing
China

Xiaoqun Chen
China Academy of Space Technology
Beijing
China

Editorial Office
MDPI
St. Alban-Anlage 66
4052 Basel, Switzerland

This is a reprint of articles from the Special Issue published online in the open access journal *Electronics* (ISSN 2079-9292) (available at: https://www.mdpi.com/journal/electronics/special_issues/8L6C28OI40).

For citation purposes, cite each article independently as indicated on the article page online and as indicated below:

Lastname, A.A.; Lastname, B.B. Article Title. *Journal Name* **Year**, *Volume Number*, Page Range.

ISBN 978-3-7258-1309-4 (Hbk)
ISBN 978-3-7258-1310-0 (PDF)
doi.org/10.3390/books978-3-7258-1310-0

Cover image courtesy of Le Xu

Contents

About the Editors

Le Xu

Le Xu received B. S. and M. Eng. degrees in in Microwave Telecommunication Engineering in 2002 and 2005, respectively, before receiving a Ph.D. degree in Radio Physics, all from Xidian University, China. From 2013 to 2014, he was a visiting scholar at the Poly-Grames Research Center at the École Polytechnique (University of Montréal), Canada. Dr. Xu is currently an Assistant Professor at the School of Electronic Engineering, Xidian University, instructing several courses for under-graduate students, such as Electromagnetic Fields and Microwave Technology. His research is in the area of computational electromagnetics. He focuses on the Physical Optics Method and Finite-Difference Time-Domain Method. He is also interested in the application of computational electromagnetic techniques to problems in engineering.

Xiaowei Shi

Xiaowei Shi received his B.Sc. degree in Physics in 1982, his M. Eng. degree in Electrical Engineering in 1990, and his Ph.D. degree in Electromagnetic Field and Microwave Technology in 1995, respectively, all from Xidian University, Xi'an, China. Between 1996 and 1997, he worked with the Electronics and Telecommunications Research Institute of Korea for his postdoctoral research work. He has been a professor and Ph.D. student advisor at Xidian University. His research interests include the theory of microwave network, microwave measurement, electromagnetic inverse scattering, the theory of electromagnetic variation, electromagnetic compatibility, and smart antenna technology (which has become his main area of interest in recent years). He has published more than 200 papers about the above research areas.

Rui Li

Rui Li received a B. Eng. degree in Electronic Engineering and M. Eng. and Ph. D degrees in Electromagnetic Field and Microwave Technology from Xidian University, China, in 2003, 2006 and 2011, respectively. She is currently a lecturer in the Department of Telecommunication Engineering at Xidian University. Her research interests include, among others, evolutionary computation, adaptive filter, smart antenna techniques, and pattern synthesis.

Xiaojie Chen

Xiaojie Chen received B.S., M.S., and Ph.D. degrees from Xidian University, Xi'an, China, in 2003, 2006, and 2008, respectively, all in Electromagnetic Field and Microwave Technology. From December of 2008 to January of 2011, she was a post-doctoral research fellow with the School of Electronic and Information Engineering at Beihang University, Beijing, China. From March of 2011 to March 2014, she worked with the 54th Institute of China Electrics Technology Group, Shijiazhuang, China. In April of 2014, she joined the CAEP Software Center for High Performance Numerical Simulation, Beijing, where she is currently a senior engineer. Her primary interest is in computational electromagnetics, especially the High-Frequency Approximate Method, Finite-Difference Time-Domain Method, and the Moment Method. Her research interests also include electromagnetic compatibility and antenna technology.

Xiaoqun Chen

Xiaoqun Chen, a researcher, graduated from the Xi'an University of Electronic Science and Technology with a doctoral degree in Electronic Science and Technology. He works at the China Academy of Space Technology, mainly focusing on advanced communication satellite payload technology and communication satellite system design. Since beginning his career, he has published more than 60 articles, conducted over 20 SCI searches, granted more than 10 patents, and authored four national military standards and four aerospace industry standards.

electronics MDPI

Editorial

Application of Computational Electromagnetics Techniques and Artificial Intelligence in the Engineering

Rui Li * and Le Xu

School of Electronic Engineering, Xidian University, Xi'an 710071, China; lexu@mail.xidian.edu.cn
* Correspondence: ruili@mail.xidian.edu.cn

Since the establishment of Maxwell's equations in the 19th century, computational electromagnetics has undergone a century of stable development [1]. Numerical algorithms such as the finite difference method, the finite element method, the moment method, and the high-frequency approximation method have been developed, serving as important cornerstones in the field of modern electronics and information [2,3].

However, traditional computational electromagnetic algorithms, when faced with device optimization design scenarios that require a large number of repeated calculations, cannot effectively utilize historical data to mine the underlying physical laws; therefore, they cannot achieve faster iteration [4]. At the same time, traditional algorithms generate a large number of intermediate calculation results and cached data when computing large-scale simulation tasks, resulting in the wastage of computing resources [5].

Artificial intelligence algorithms, on the other hand, can automatically extract non-linear mapping relationships from electromagnetic simulation parameters to the required electromagnetic response through learning and training. This approach fully utilizes historical data information to improve the simulation efficiency, advancing intelligent electromagnetic computing [6].

Moreover, AI can enhance the efficiency and accuracy of reverse optimization tasks by introducing prior information on geometric features or other electromagnetic characteristics of optimization targets. This has achieved many representative results in applications such as antenna structure optimization, holographic imaging design, and inverse scattering imaging [7].

In recent years, a variety of optimization algorithms, such as the particle swarm optimization (PSO), the ant colony optimization (ACO), and differential evolution (DE), have received widespread attention. PSO, ACO, DE, and population-based incremental learning (PBIL) have gained recognition in filter design. A novel method merging MOEA/D and a 1D-CAE network structure is proposed to design microwave filters (Contribution 1). The 1D-CAE network structure replaces the traditional full-wave simulation, which greatly reduces the design time of the filter.

The design of a hybrid location service strategy is introduced (Contribution 2), including forwarding strategy selection, switching design, and forwarding strategy design, as well as the data packet transmission process based on a reactive routing protocol. At the same time, the network problems of small unmanned aerial vehicle systems (UAS), especially the routing problem of flight self-organizing networks (FANETs), are summarized, and the research status and future development directions of related routing strategies are discussed.

An improved combination method is presented (Contribution 3), which combines the Method of Moments (MoM) and Consistent Diffraction Theory (UTD) to calculate the radiation characteristics of antenna arrays installed on large electrical platforms. This method first sequentially feeds each array unit in a MoM processor and then inputs the results into the UTD process to calculate the perturbation pattern. When solving the current coefficient in MoM, the principle of linear superposition is used to achieve separate feeding

Citation: Li, R.; Xu, L. Application of Computational Electromagnetics Techniques and Artificial Intelligence in the Engineering. *Electronics* **2024**, *13*, 1835. https://doi.org/10.3390/electronics13101835

Received: 18 April 2024
Accepted: 19 April 2024
Published: 9 May 2024

for each array unit. This method can avoid the problem of unreliable UTD results when the equivalent phase center of the antenna is obscured and reduce the root mean square error from over 15 dB to about 1 dB in the given T-shaped example. Furthermore, this method does not require a significant amount of computing resources.

The effect of introducing a collector ring on the grounding loss resistance of a very low frequency (VLF) umbrella transmitting antenna grounding system is investigated (Contribution 4). By establishing a geometric deconstruction model of the collector ring, a VLF umbrella transmitting antenna model, an umbrella antenna ground network system model, and an average conductivity formula, the working principle of the collector ring is analyzed. The simulation of the grounding loss resistance of collector ring structures with different layers and degrees of damage using Feko software is described. The simulation results show that, in the case of intact and undamaged ground grid system, increasing the number of collector rings has a relatively weak impact on the ground loss resistance. However, when the local network conductor breaks, the collector ring structure can effectively reduce the grounding loss resistance.

An improved diode SPICE model based on measurement and nonlinear fitting stochastic optimization algorithm is proposed (Contribution 5). By analyzing the electromagnetic interference mechanism and characteristic parameters of the diodes in detail, the model accuracy is successfully improved. The experimental results show that the model can truly reflect the characteristics of electromagnetic interference, ensuring that the harmonic components and amplitudes are within three times the fundamental frequency and verifying the accuracy of the electromagnetic compatibility model.

For solving high computational costs in filter optimization problems, an adaptive online updated 1D convolutional autoencoder (AOU-1D-CAE) surrogate model is proposed (Contribution 6). This model can automatically collect, select, and use data as training samples, and as the number of samples increases, the model's learning ability gradually strengthens. The experimental results indicate that the AOU-1D-CAE improves the data collection efficiency and enhances the predictive performance. Combined with Particle Swarm Optimization (PSO), this surrogate model can assist PSO in finding the global optimal solution, avoid falling into local optima, and improve optimization efficiency. The effectiveness of the proposed surrogate model is verified using examples of two cavity filters.

The finite difference time domain method is adopted to comprehensively analyze the impact of front and rear door coupling on electronic devices, considering the coupling scheme of the voltage source injection and plane wave irradiation occurring simultaneously, and introducing an improved equivalent circuit model based on physical models (Contribution 7). Taking PIN limiters as an example, the authors analyze their performance impact and propose suggestions for protection.

A method to measure the shielding effectiveness of irregular cavities is proposed (Contribution 8). An electromagnetic topology model is established for the cavity, simulating its field distribution characteristics. Combined with testing methods for regular cavities, a comprehensive approach to characterize and test the shielding effectiveness of irregular cavities is presented, effectively verifying its accuracy.

A novel tuning fork-shaped tri-band planar antenna (NTTPA) is introduced for multiple wireless communication bands (Contribution 9). Its asymmetrical design generates three operating bands, covering LTE, WLAN, and WiMax frequencies. Fabricated on FR4 with a compact size, it offers a wide bandwidth and stable radiation pattern. Ideal for wireless systems, it is cost-effective and easy to process.

In another article, artificial neural networks are used to predict the initial design of multiband planar antennas (Contribution 10). Trained on normalized antennas analyzed through modal analysis, the networks quickly select suitable geometries for desired operating bands. This neural pre-design concept simplifies and accelerates antenna design, promising to refine and enhance future multiband antenna designs.

A balanced bandpass filter is proposed (Contribution 11), which utilizes two half wavelength stepped impedance resonators to achieve a differential mode passband, with improved frequency selectivity, high common mode suppression, and wide stopband characteristics. The filter adopts a balanced U-shaped microstrip slot line transition structure for feeding, reducing design complexity. The simulation and measurement results are consistent, verifying the effectiveness of the design.

The design of a balanced bandpass filter based on slot line short cut loaded resonators is presented (Contribution 12); it has a wide differential mode bandwidth, steep passband selectivity, and high common mode suppression characteristics. By adjusting the size and gap of the resonator, the passband performance can be flexibly controlled. It adopts a U-shaped microstrip line to the slot line transition structure to achieve broadband common mode suppression and simplify design. The experimental verification shows that the performance of the filter is good, and the simulation and measurement results are consistent.

Conflicts of Interest: The authors declare no conflict of interest.

List of Contributions

1. Wei, Y.; Qi, G.; Wang, Y.; Yan, N.; Zhang, Y.; Feng, L. Efficient Microwave Filter Design by a Surrogate-Model-Assisted Decomposition-Based Multi-Objective Evolutionary Algorithm. *Electronics* **2022**, *11*, 3309.
2. Xin, N.; Chen, X.; Chen, T.; Liu, F.; Liu, K.; Lin, Y. Research on Multi-Strategy Routing Protocol in Flying Ad Hoc Networks. *Electronics* **2022**, *11*, 3327.
3. Wang, N.; Liu, J.; Zhang, Q.; Zhang, Y. An Improved Combination Method of MoM and UTD for Calculating the Radiation Characteristics of Antenna Arrays Mounted around Electrically Large Platform. *Electronics* **2022**, *11*, 3369.
4. Quan, Y.; Xie, H.; Yang, C.; Yu, H.; Liu, X. Effects of Introducing Confluence Rings on Ground Loss Resistance of VLF Umbrella-Type Antenna. *Electronics* **2022**, *11*, 3419.
5. Ren, D.; Xu, G.; Li, J.; Pan, Z.; Zhao, X.; Du, P. Improvement of the SPICE Model of Diode Based on Measurement and Nonlinear Fitting Random Optimization Algorithm. *Electronics* **2022**, *11*, 3461.
6. Zhang, Y.; Wang, X.; Wang, Y.; Yan, N.; Feng, L.; Zhang, L. Accurate Design of Microwave Filter Based on Surrogate Model-Assisted Evolutionary Algorithm. *Electronics* **2022**, *11*, 3705.
7. Liu, T.; Xu, L.; Li, Q.; Yao, B.; Shi, X. Joint Analysis of Front-Door and Back-Door Couplings of PIN Limiter Based on Improved Equivalent Circuit Model. *Electronics* **2022**, *11*, 3921.
8. Ren, J.; Pan, Y.; Zhou, Z.; Zhang, T. Research on Testing Method for Shielding Effectiveness of Irregular Cavity Based on Field Distribution Characteristics. *Electronics* **2023**, *12*, 1035.
9. Li, Q.; Fang, J.; Ding, J.; Cao, W.; Sun, J.; Guo, C.; Liu, T. A Novel Tuning Fork-Shaped Tri-Band Planar Antenna for Wireless Applications. *Electronics* **2023**, *12*, 1081.
10. Lahiani, M.; Raida, Z.; Veselý, J.; Olivová, J. Pre-Design of Multi-Band Planar Antennas by Artificial Neural Networks. *Electronics* **2023**, *12*, 1345.
11. Li, Q.; Fang, J.; Cao, W.; Sun, J.; Ding, J.; Tie, W.; Wei, F.; Zhai, C.; Wu, J. Optimization and Design of Balanced BPF Based on Mixed Electric and Magnetic Couplings. *Electronics* **2023**, *12*, 2125.
12. Liu, Y.; Wu, W.; Li, J.; Zhao, M.; Wei, F. A Balanced BPF with Wide Bandwidth and Steep Selectivity Based on Slotline Stub Loaded Resonators (SSLRs). *Electronics* **2023**, *12*, 3389.

References

1. Smith, J.C. On the electromagnetic effects due to variations in the density of electric currents. *Philos. Trans. R. Soc. Lond.* **1873**, *163*, 375–424.
2. Jones, D.S. The finite element method in electromagnetics. *J. Phys. D Appl. Phys.* **1993**, *26*, 237.
3. Chen, Z. The finite element method and its application in electromagnetics. *J. Comput. Appl. Math.* **2002**, *149*, 1–9.

4. Wang, L.; Zhang, Y.; Liu, H. Accelerating electromagnetic simulation with machine learning. *IEEE Trans. Antennas Propag.* **2018**, *66*, 7105–7112.
5. Li, Y.; Zhang, H. Resource-efficient electromagnetic simulation using deep learning. *IEEE Trans. Microw. Theory Tech.* **2019**, *67*, 2836–2845.
6. Liu, J.; Chen, X.; Wang, P. Intelligent electromagnetic computing: A new frontier in electromagnetic simulation. *IEEE Antennas Propag. Mag.* **2022**, *64*, 6–17.
7. Guo, X.; Yang, F. Enhanced reverse optimization using artificial intelligence in electromagnetic design. *IEEE Trans. Electromagn. Compat.* **2021**, *63*, 921–930.

electronics

MDPI

Article

Efficient Microwave Filter Design by a Surrogate-Model-Assisted Decomposition-Based Multi-Objective Evolutionary Algorithm

Yongfeng Wei [1], Guangfei Qi [1], Yanxing Wang [1], Ningchaoran Yan [1], Yongliang Zhang [2,*] and Linping Feng [3,4]

1 School of Electronics and Information Engineering, Inner Mongolia University, Hohhot 010021, China
2 School of Transportation Institute, Inner Mongolia University, Hohhot 010021, China
3 Shaanxi Key Laboratory of Deep Space Exploration Intelligent Information Technology, School of Electronics and Information Engineering, Xian Jiaotong University, Xi'an 710049, China
4 School of Microelectronics, Xian Jiaotong University, Xi'an 710049, China
* Correspondence: namar@imu.edu.cn; Tel.: +181-4834-2159

Abstract: As a crucial frequency selection device in modern communication systems, the microwave filter plays an increasingly prominent role. There is a great demand for the multi-objective design of microwave filters. The filter's performance affects the quality of the whole communication system directly. However, traditional multi-objective electromagnetic (EM) optimization design demands repetitive EM simulations to adjust the physical parameters of the microwave filters. Accordingly, using electromagnetic simulation directly to design optimization is quite expensive. Given this situation, this paper applies a novel surrogate model based on one-dimensional convolutional autoencoders (1D-CAE) into the multi-objective algorithm evolutionarily based on decomposition (MOEA/D) for the first time. This approach uses MOEA/D as the multi-objective optimizer, and a novel low-complexity surrogate model based on one-dimensional convolutional autoencoders (1D-CAE) is constructed to predict the expensive EM simulation results. The surrogate model based on 1D-CAE is used to generate the results of scalar subproblems of MOEA/D, which greatly improves the design efficiency. Compared with the traditional design methods based on an EM solver, this method not only effectively optimizes multiple design objectives but also completes the design of microwave filters in a shorter time. The proposed method is verified using the design of a sixth-order ceramic filter and a seventh-order metal cavity filter.

Keywords: multi-objective optimization; microwave filter design; MOEA/D; surrogate model; one-dimensional convolutional autoencoders

Citation: Wei, Y.; Qi, G.; Wang, Y.; Yan, N.; Zhang, Y.; Feng, L. Efficient Microwave Filter Design by a Surrogate-Model-Assisted Decomposition-Based Multi-Objective Evolutionary Algorithm. *Electronics* **2022**, *11*, 3309. https://doi.org/10.3390/electronics11203309

Academic Editor: Leonardo Pantoli

Received: 13 September 2022
Accepted: 4 October 2022
Published: 14 October 2022

1. Introduction

The microwave filter is quite a vital RF device in wireless communication systems. With the rapid development of wireless communication technology, it is very important to improve the efficiency of filter design. In recent years, a variety of optimization algorithms, such as particle swarm optimization (PSO), ant colony optimization (ACO), and differential evolution (DE), are receiving widespread attention. PSO, ACO, DE, and population-based incremental learning (PBIL) have gained recognition in filter design [1–6]. However, in the design process of microwave filters, engineers cannot just consider a single design objective but need to achieve equilibrium among multiple different objectives, so as to make microwave filters meet the prescribed requirement. For example, in the wideband balun bandpass filters (BPFs), the phase imbalance, magnitude imbalance, and frequency responses are important indicators to determine the filter performance. If any response parameter fails to meet the prescribed requirement, it will directly affect the final performance of the filter balun structure [7]. Therefore, multi-objective optimization is gradually replacing single-objective optimization and has become the mainstream optimization form of modern microwave filter design [8–10].

For the multi-objective optimization method, MOEA/D [11] is one of the most commonly used methods for multi-objective optimization problems. MOEA/D uses the aggregation function to disassemble multi-objective problems into several scalar optimization subproblems with neighborhood relations and to co-evolve rather than relying on Pareto domination. Due to the decomposition operation, MOEA/D has prominent advantages to maintain the distribution of solutions. Several MOEA/D variants have been proposed to improve MOEA/D. A variant called MOEA/D-ACDP is proposed in [12], which improves the population diversity in infeasible regions through the angular information of the objective function. Qi et al. proposed a variant that can adaptively adjust weights (MOEA/D-AWA) [13], which initializes and adjusts the weight vector in a new way. A form of fusion (MOEA/DD) that combines decomposition and domination is proposed in [14]. Li et al. proposed MOEA/D-STM [15], which utilizes a matching model to coordinate the selection process in MOEA/D to balance convergence and diversity.

In addition to a better optimization algorithm, a more important problem is how to improve the efficiency of multi-objective microwave filter design. Recently, EM-based optimization methods were developed into a vital optimization approach for microwave filter design. The meshes applied to EM simulations are usually produced by mesh adaptation methods. For traditional coarse- and fine-mesh SM, as long as the geometric parameter value changes, it is necessary to regenerate the coarse mesh to adapt to this change. However, in the optimization design of microwave filters, a great deal of EM simulations is required to acquire the optimal design space parameters commonly. Therefore, the traditional direct EM-based microwave filter design takes a long time to meet design requirements. Space mapping (SM) is the key technology to solving the above problem. SM has shown great usability in computer-aided optimization design [16–19]. The SM concept combines the computational efficiency of coarse models with the accuracy of fine models [20]. Although the fine models are accurate, they may be expensive. The SM technique constructs a mathematical relation between the coarse and the fine models, orients a large number of CPU-intensive computations to the coarse model, and preserves the accuracy offered by the fine model [21]. At present, the maturity of applying SM to microwave component optimization design is increasing day by day [22–25].

In recent years, in order to improve the efficiency of filter design, many optimization methods based on SM have been proposed. Still, these methods only consider specific types of filter structures [26]. A surrogate-model-assisted PSO algorithm is proposed in [27]. The algorithm can effectively shorten the design time, but it can only be used for single-objective microwave filter design. Hence, most of the available methods cannot combine better optimizers and shorter design time at the same time. In [28], a novel off-line surrogate model is proposed to design filters with more than 10 variables. More than 2000 samples are used to construct this single high-accuracy surrogate model [28]. However, this is unrealistic for more complex filters, and most filter design/optimization methods try to avoid such a step.

To address the above problems, a novel microwave filter design technique is presented, called MOEA/D based on 1D-CAE. Convolutional autoencoder (CAE) is an important type of deep learning model that is widely applied into various fields, such as image denoising, neural style transfer, and so on. Masci et al. [29] developed CAE to process 3D image data. In [30], a deep CAE is proposed to process high-resolution SAR images. Combining stack and CAE, [31] put forward a new type of structure of low-light image enhancement. In [32], a three-layer CAE architecture and an effective algorithm are designed to learn CAE parameters for device-free localization. However, CAE, as a surrogate model, is introduced into the research of microwave filter design, which is still blank. Only [27] has made efforts in this regard, but it cannot complete the design of a multi-objective microwave filter. Therefore, in this paper, CAE and MOEA/D are combined for the first time to realize the fast multi-objective design of microwave filters.

In this paper, the 1D-CAE network structure predicts the filter's S-parameters at potentially better sampling points by building an inexpensive surrogate model. Autoencoder is an artificial neural network that learns the efficient representation of input data. However, in the traditional autoencoder training process, it is very easy to copy the input to the output, which leads to the poor predictive ability of the trained model. The model's prediction ability will be greatly reduced when the input samples are too complex. For filter design, the relationship between the physical parameters of the filter and the S-parameters is complex. Therefore, it is difficult for autoencoders to extract effective features from filter data. The convolutional neural network (CNN) can perceive adjacent dimensions of data and obtain local features through receptive fields and parameter sharing. This process is based on convolution of the kernel. A variety of local features can be obtained by constructing data samples with multiple convolution kernels. In addition, the pooling layer can reduce the parameters in the whole neural network. Therefore, iterative convolution and pooling can be utilized to extract the final features of the data samples. The framework of 1D-CAE is as follows. In the encoder part, features are extracted by CNN, which continuously iterates the convolution and pooling of multiple convolution kernels to reduce the number of features. In the decoder part, the extracted features are used to reconstruct the sample data by the reshape operation and full connection layer. It can be said that CAE combines the advantages of CNN and AE in feature learning and data reconstruction.

In this paper, the 1D-CAE network structure uses the one-dimensional convolutional neural network (1D-CNN) as an encoder for feature extraction. That is, the relationship between a microwave filter's physical parameters and responses is established. After fully learning by the encoder, the physical parameters are predicted into the characteristic response by the decoder directly. As shown in Figure 1, the decoding part of 1D-CAE is applied to construct a novel surrogate model to displace the full-wave EM simulation, which greatly shortens the calculation time of MOEA/D. This paper uses the loss function to determine whether the 1D-CAE prediction is accurate. The MOEA/D algorithm based on the 1D-CAE network structure aims to:

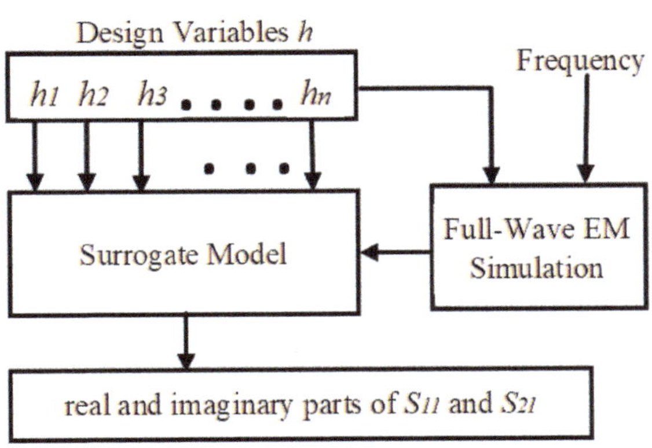

Figure 1. Surrogate model constructed by 1D-CAE.

(1) Multiple design objectives can be jointly optimized;

(2) Obtain comparable results as the traditional multi-objective design approach (directly using EM solver for MOEA/D);

(3) Finish the multi-objective design of the filter in a shorter time (a few hours to one day or so);

(4) When the specific characteristics of the filter are not considered, it can be universally used in most types of filters.

The remainder of this article is structured as follows. Section 2 discusses related works and methods, illustrates the basic theory of 1D-CAE network structure and the MOEA/D algorithm in terms of mathematical formulation, and introduces a detailed framework of 1D-CAE-based MOEA/D. In Section 3, the practicability and effectiveness of the presented technology are confirmed by two practical microwave filter design examples. Section 4 concludes the paper.

2. Materials and Methods

2.1. D-CAE Network Structure

2.1.1. Autoencoder's Framework

Autoencoder is an artificial neural network that learns the efficient representation of input data. It is trained to try to reproduce input to output. The dimension of the encoder part is universally much smaller than that of the input data. So, AE can be applied for dimensionality reduction. By extracting features, AE also has the function of generating model, that is, it can randomly generate data similar to training data.

As shown in Figure 2, the typical structure of AE has two parts (encoder part and decoder part). The encoder part is composed of an input layer and a hidden layer, and it is used to extract latent features from original inputs. The original input data are $X = (x_1, \ldots, x_n)$, and the function of an encoder is to map the input sample into the latent space as the feature h:

$$h = Encoder(X), \quad (1)$$

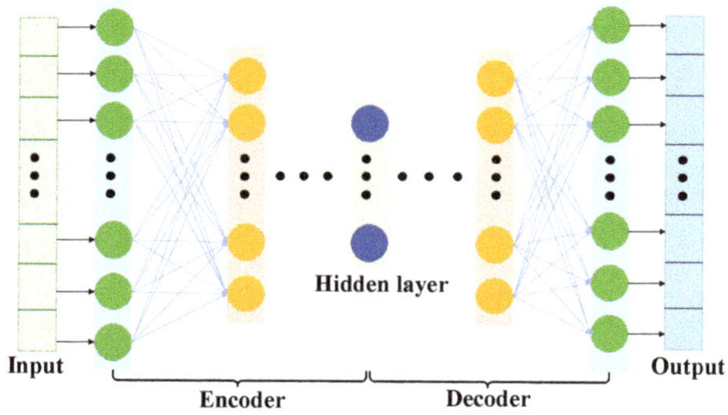

Figure 2. The typical structure of AE.

The decoder part is composed of a hidden layer and an output layer. To obtain a decoded output vector $X*$, the decoder part maps the feature h to the original input space:

$$X* = Decoder(h), \quad (2)$$

AE is committed to infinitely narrowing the gap between input samples and output samples. This gap is represented by the loss function:

$$L = K(X, X*), \quad (3)$$

2.1.2. One-Dimensional Convolutional Autoencoders

The traditional autoencoder can easily copy the input sample to the output sample. However, the scope of application of this model is limited. The model's prediction ability will be greatly reduced when the input samples are too complex. For filter design, the relationship between the physical parameters of the filter and the S-parameters is complex. In this paper, the 1D-CAE network structure based on AE and 1D-CNN is adopted to build the surrogate model. Including three convolution layers, three pooling layers, a reshape module, and a full connection layer, namely conv1, pooling1, conv2, pooling2, conv3, pooling3, reshape1, and full-con1, the encoder module is used to encode the input data. The decoder module consists of the reshape operation and full connection layer, namely full-con2, full-con3, full-con4, and reshape2. Figure 3 is an internal detailed structure diagram of 1D-CAE. Different types of layers are described in detail below.

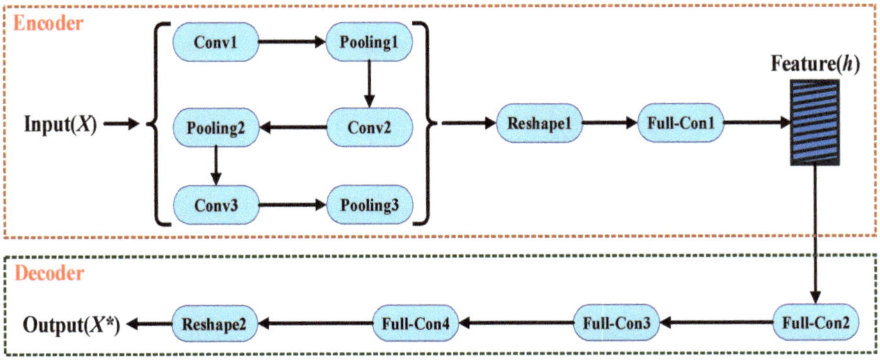

Figure 3. The detailed structure of the 1D-CAE network structure.

As a feature extractor, the convolution layer makes use of several convolution kernels for convolution calculation to obtain more features. Let f_z indicate the output of the *Z-th* kernel. f_z is determined as follows [33]:

$$f_z = \text{ReLU}(\sum X \odot \omega_z + b_z), \tag{4}$$

$$\text{ReLU}(x) = \max(0, x), \tag{5}$$

where $\odot$ represents convolution operation, ω_z is the *Z-th* convolution kernel, and b_z is the offset corresponding to the *Z-th* convolution kernel. Z indicates the number of channels. ReLU is a rectifier linear unit.

The pooling layer can reduce the parameters in the whole neural network. The common types of pooling layers are max pooling and average pooling. This paper adopts max pooling. Max pooling only retains the most significant features so that it can reduce the size of the training sample and improve computational efficiency. For the k-length feature of the convolution layer in the *Z-th* channel, the output of the pooling layer is determined by the following formula [33]:

$$P_z(n) = \max_{0 \leq n \leq \frac{k}{st}} \{f_z(nS, (n+1)S)\}, \tag{6}$$

where f_z represents the input. S is the size of the pooling window, and st is the step size.

The reshape module transforms the extracted feature vector into a one-dimensional form to facilitate coding in the full connection layer. The calculation process is as follows [34]:

$$F = \sin\left(w_f * E + b_f\right), \tag{7}$$

where E is the eigenvector output by the last level pooling layer, w_f is the weight of the full connection layer, b_f indicates the offset of a full connection layer, and sin represents a kind of activation function.

2.1.3. 1D-CAE Network Structure Introduction in Filter Design

For the encoding part of the 1D-CAE, the input is the real part and imaginary part of the filters *S11* and *S21*, and the output is the geometric parameters of the filter to be optimized. On the contrary, the input of the decoding part is the geometric parameters to be optimized, and the real and imaginary parts of *S11* and *S21* are their outputs. Therefore, the whole 1D-CAE network structure is a system in which the input part is the real *S11* and *S21* and the output part is the predicted *S11* and *S21*, that is, the generation model function of the 1D-CAE network structure.

For the actual optimization of this paper, the performance of the 1D-CAE network structure is reflected by the combination of rebuild loss (L_r) and forecast loss (L_f). Let us use rebuild loss to assess the performance of 1D-CAE network structure reconstruction filter features. Rebuild loss is determined by the following formula [27]:

$$L_r = \frac{1}{M}\sum_{i=1}^{M}|S_{em} - S_{pr}|, \tag{8}$$

where M represents the total training data, S_{em} represents the simulated S-parameters, and S_{pr} is the S-parameters of the reconstructed 1D-CAE network structure.

The predicted losses are expressed as follows [27]:

$$L_f = \frac{1}{M}\sum_{i=1}^{M}|H_{em} - H_{pr}|, \tag{9}$$

where M represents the total training data, H_{em} represents the realistic value of the physical parameters, and H_{pr} is the predicted physical parameters' value.

The total loss function is determined by the following formula [27]:

$$L_t = L_r + kL_f, \tag{10}$$

where k is a regularization parameter. So as to make the network prediction more accurate, it is essential to minimize the value of the loss function (L_t). In this paper, the Adam optimizer is used to optimize the loss function of the network.

2.2. MOEA/D Algorithm

2.2.1. Multi-Objective Optimization Problem

The multi-objective optimization problem (MOPs) makes multiple objectives as best as possible in a given region at the same time. It is expressed by the following formula:

$$\text{Minimize} f(x) = (f_1(x), \ldots, f_M(x)) \quad \text{subject to } x \in \text{X}, \tag{11}$$

where $x = (x_1, \ldots, x_e)$ represents the e-dimensional vector to be solved, X represents the boundary range of x, $f_i(x)$ represents the *i-th* objective function to be optimized, and M is the dimension of the goal vector. MOPs cannot receive a single solution that can meet all objectives concurrently. Accordingly, the solution of MOPs is usually a set of equilibrium solutions. Suppose x^E and x^F are two solutions of the MOPs in (11); x^E is said to Pareto dominate x^F if and only if $f_i(x^E) \leq f_i(x^F)$ for all $i \in \{1, 2, \ldots, M\}$ and $f_j(x^E) \leq f_j(x^F)$ for at least one $j \in \{1, 2, \ldots, M\}$. A solution x^E is Pareto optimal when there is no $x^G \in \text{X}$ that Pareto dominates x^E. The Pareto optimal set consists of all Pareto optimal solutions, and the Pareto front (PF) is the projection of the Pareto optimal set in the objective space.

2.2.2. MOEA/D

MOEA/D disassembles the MOPs into several single-objective optimization subproblems with neighborhood relations, and each subproblem is a dissimilar set of all objectives. Then, by analyzing the information of adjacent problems, all subproblems are optimized concurrently by an evolutionary algorithm. Due to the decomposition operation, this method can well maintain the distribution of solutions. In addition, by analyzing the information of adjacent problems to optimize, it is effective to avoid falling into local optimum. The detailed introduction to the important part of the MOEA/D is as follows.

I. Weight vector

In order to disassemble the MOPs into several single-objective subproblems, MOEA/D needs to distribute even weight vectors in the target space. The number of weight vectors is the same as the population size. If the population size is N, the number of weight vectors is N. Each weight vector turns the MOPs into a single-objective problem. N sets of weight vectors are N single-objective optimization problems and$w_N = (w_{N1}, w_{N2}, \ldots, w_{NM})$ is the *N-th* weight vector, so:

$$w_{N1} + w_{N2} + \ldots + w_{NM} = 1, \tag{12}$$

where M represents the number of problems to be optimized. In addition, the uniformity of weight vector distribution should be as good as possible to improve the accuracy of optimization results.

II. Decomposition strategy

The MOEA/D aims to convert the multi-objective optimization problem to single-objective optimization problem through an aggregation function. The generally applied decomposition strategies include penalty-based boundary intersection (PBI), weighted sum (WS), and Tchebycheff (TCH), among which the WS cannot solve nonconvex functions. The Tchebycheff method is used in this paper. The formula of Tchebycheff method is expressed by the following formula:

$$\text{Minimize } g^{\text{TCH}}(x|\, w, z^*) = \max_{i=1,2,\ldots,M} \{w_i \cdot |z_i^* - f_i(x)|\}, \tag{13}$$

where z is the optimization result to be achieved (the reference point). Take double objectives as an example, calculating $w_1 \cdot |z_1^* - f_1(x)|$ and $w_2 \cdot |z_2^* - f_2(x)|$, respectively, to take the maximum value. The larger the value, the farther away from the reference point on this objective function. Supposing $w_1 \cdot |z_1^* - f_1(x)|$ is larger, please gradually change x so that this value is closer and closer to z^* until it reaches the corresponding point on Pareto front. This process is actually finding the minimum value of the function $g^{\text{TCH}}(x) = w_1 \cdot |z_1^* - f_1(x)|$. If $w_1 \cdot |z_1^* - f_1(x)|$ reaches its minimum value, $w_2 \cdot |z_2^* - f_2(x)|$will also reach its minimum value. This is true for weight vector w. Each weight vector obtains the corresponding solution in this way.

III. Neighborhood structures

Neighborhood structure is determined by the Euclidean distance among the weight vectors, which is the key to generating a new solution. The MOEA/D assumes that the solutions on adjacent weight vectors are similar, and each weight vector has neighbors. After generating a new solution, compare the newly generated solution with all solutions in its neighborhood (including the current solution of the current subproblem). When the newly generated solution is better, all poor neighbors will be displaced by the newly generated solution.

2.3. Filter Design by Surrogate-Modeling-Assisted MOEA/D

In this paper, the samples generated by full-wave EM simulation are used to train a 1D-CAE network to obtain the microwave filter responses in a shorter time in the design space. MOEA/D is applied to calculate the geometric parameters of the filter that meet the design requirements. The complete process of microwave filter design by 1D-CAE-network-assisted MOEA/D is given in Figure 4. In this process, there are several key points that need to be described in detail:

(a) Step 2: Data collection step. The purpose of data acquisition is to collect a certain number of training and test sets to develop a surrogate model. A set of data sets consists of a set of geometric parameters of the filter and its corresponding S-parameters. Electromagnetic simulations were performed by a high frequency structure simulator (HFSS) to obtain training and test sets. The values of the geometric parameters (H) are determined by the following formulas:

$$H = H_0 + kH_0, \tag{14}$$

where H_0 is the initial geometric parameter of the filter (center point), k is a random number in the data collection range, and this range is around the center point and is determined considering the characteristics of the filter in this example.

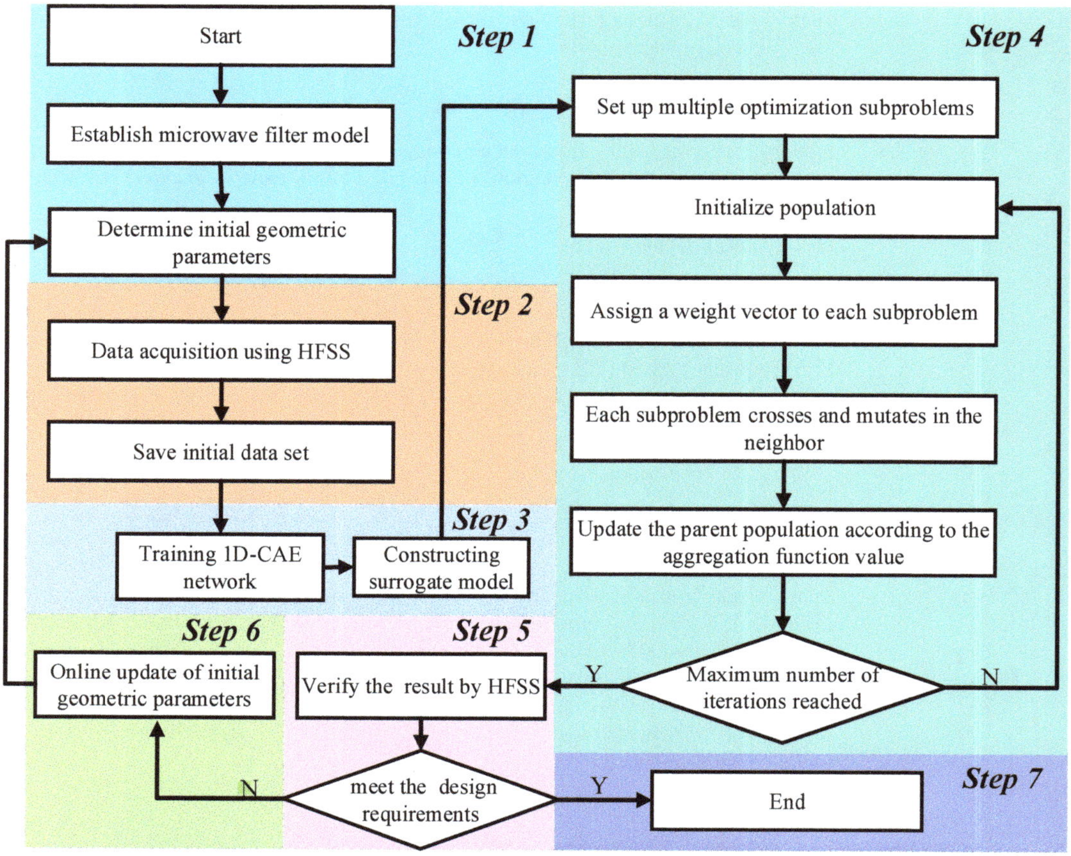

Figure 4. The complete flow diagram of the filter design technique.

Next, we input H_0, k, the number of data acquisitions and the center frequency of the filter, and so on into the data acquisition system. The data acquisition system will automatically drive the HFSS to complete the acquisition process until the preset number of data acquisitions is reached and the system stops running. The entire process requires no human intervention.

(b) Step 3: Surrogate model construction step. In this step, the data set generated in step 2 is divided into test set and training set according to 1:5. The value of the loss function is calculated by Equation (10). The Adam optimizer is used to optimize the loss function of the network.

The complete steps of 1D-CAE-network-aided MOEA/D to design filter are as follows:

Input:

⟨1⟩ H_0,k, the number of data acquisitions;

⟨2⟩ A multi-objective filter design problem with m objectives;

⟨3⟩ Termination conditions (e.g., filter design goal, maximum number of iterations);

⟨4⟩ MOEA/D parameters: the number of sub-problems (N); the number of neighbors of each weight vector (T); the population size; uniformly distributed weight vector.

Output:

⟨1⟩ Filter geometry parameter values that meet design requirements.

Step 1: Determine a model step.

Step 1.1: Establish a microwave filter model.

Step 1.2: Determine the initial geometric parameters of the microwave filter, that is, H_0 in Equation (14).

Step 2: Data collection step.

Step 2.1: The geometric parameter value H is obtained according to Equation (14), and then the response (S parameter) of this microwave filter is generated by HFSS simulation.

Step 2.2: Save the S-parameters in the form of the real part and the imaginary part.

Step 3: Surrogate model construction step.

Step 3.1: Training 1D-CAE network structure. The value of the loss function is calculated by Equation (10). The Adam optimizer is used to optimize the loss function of the network.

Step 3.2: The surrogate model is constructed via the trained 1D-CAE network structure.

Step 4: MOEA/D step.

Step 4.1: Based on the multiple design requirements of the microwave filter, the corresponding optimization subproblems are set up.

Step 4.2: Initialize population.

Step 4.2.1: Create an external population (EP) to store outstanding individuals, initially empty.

Step 4.2.2: Calculate the Euclidean distance between any two weight vectors and find the T closest weight vectors to each weight vector. For $i = 1, 2, \ldots, N$, let $B(i) = \{i_1, \ldots, i_T\}$. $x^1, x^2, \ldots, x^N$are the nearest T weight vectors of w_j.

Step 4.2.3: Generate an initial number of random $x^1, x^2, \ldots, x^N$, using the surrogate model to calculate the fitness function value.

Step 4.2.4: Initialize $z = \{z_1, \ldots, z_M\}$.

Step 4.3: Assign a weight vector to each subproblem.

Step 4.4: Each subproblem crosses and mutates in the neighbor.

Step 4.5: Update parent population according to the aggregation function value.

For $z_j = f_j(y^{\cdot})$,

Step 4.5.1: Copy: randomly choose two indices l, q from *B(i)*, and then generate a new solution y from x^l and x^q.

Step 4.5.2: Repair: if an element of y exceeds a preset bound, its value will be reset to the max or min of the bound (generates $y^{\cdot}$ from y).

Step 4.5.3: Updatez: For eachj =1,2, . . . , m, judge whether y can replace the original extreme value. If$z_j < f_j(y^{\cdot})$, set $F(y^{\cdot})$.

Step 4.5.4: Update the domain solution *B(i)* for each weight vector $w_j$in the domain; if it is optimized, update it.

Step 4.5.5: Update EP: Remove all vectors dominated by $F(y^{\cdot})$ from EP. Add $F(y^{\cdot})$ to EP if no vector dominates $F(y^{\cdot})$ in outer population (EP).

Step 4.6: If the maximum number of iterations is reached, go to step 5. Otherwise, go to Step 4.2.

Step 5: Verification optimization result step.

Step 5.1: The optimization results obtained in step 5 are simulated with HFSS.

Step 5.2: If the results meet the design requirements, go to step 7. Otherwise, go to step 6.

Step 6: Update the geometric parameters online and then go to step 1.2 directly.

Step 7: Complete the design of a microwave filter.

3. Design Results

3.1. Sixth-Order Ceramic Filter

The first example shows the optimization design of a sixth-order ceramic bandpass filter. The structure of the sixth-order ceramic filter is shown in Figure 5. The finite transmission zeros of this microwave filter aref_1 = 2.552 GHz andf_2 = 2.858 GHz. The design geometric variables are H = $[h1, h2, h3, h4, h5, h6]^T$, and the design requirements of this filter are as follows:

- $|S11| \leq -20$ dB, for 2.6 GHz $\leq \omega \leq$ 2.8 GHz;

- $|S21| \leq -50$ dB, for $\omega = 2.552$ GHz;
- $|S21| \leq -50$ dB, for $\omega = 2.858$ GHz.

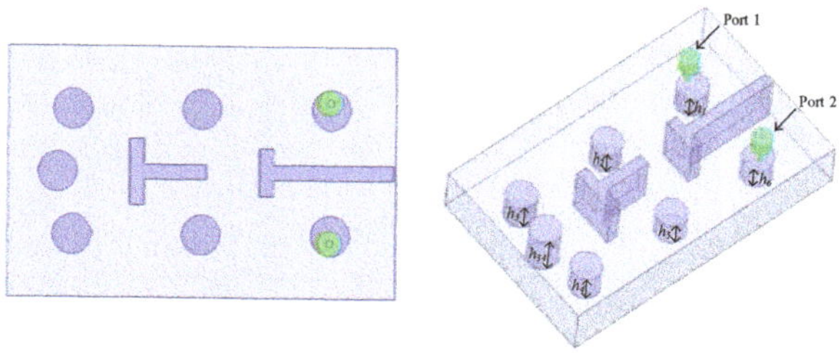

Figure 5. Structure of sixth-order ceramic filter.

The starting geometric parameter value of this example is $H = [3.4, 3.3, 3.85, 3.9, 3.4, 3.4]^T$ (mm), and its EM response indicates that the response at the starting point is far away from the design requirements.

This example consists of two main steps. The first step is to develop a surrogate model based on 1D-CAE with the real part and imaginary part of the filter's $S11$ and $S21$ as the model output and geometrical parameters as model inputs, i.e., $H = [h1, h2, h3, h4, h5, h6]^T$. The EM data are used to train the 1D-CAE. The second step is to apply the trained surrogate model to the multi-objective optimization design, and the geometric parameters can be adjusted repeatedly in the optimization process.

At first, the engineer needs to construct a 1D-CAE network structure. In Figure 6, the encoder part has a convolution layer, pooling layer (down-sampling), and full connection layer. A three-layer full connection constitutes the decoder part. The Conv1D (3, 32, 1) indicates a convolution layer. The first number (3) indicates that the convolution kernel is 3, the second number (32) indicates the number of output channels of this layer is 32 and the third number (1) indicates the stride is 1. Maxpool (3,3) represents a maximum pooling layer with a down-sampling factor value of 3 and stride size of 3. Full connection (N) represents a full connection layer with N nodes, where N is the dimension of the geometric parameters to be predicted by the filter. The input information includes real and imaginary parts of the S-parameters of 301 frequency points. After Conv1D (3, 32, 1), the input data are transformed into a matrix of 301×32 (32 channels, 301 indicates the length of a single channel). Next, the matrix became 101×32 after Maxpool (3,3).

Developing a 1D-CAE network structure requires a certain number of training sets and test sets. EM simulations are performed by a high-frequency structure simulator (HFSS) to obtain training sets and test sets. We set the range of the surrogate model to be $k = [2\%\ 2\%\ 2\%\ 2\%\ 2\%\ 2\%]^T$. This range is around the center point and is determined considering the characteristics of the filter in this example. Every time the geometric parameters change, the filter model is simulated based on this through the EM solver to gain the real part and imaginary part sample data of $S11$ and $S21$. So as to ensure the high accuracy of this model, 360 sets of data are obtained, including 288 training sample data and 72 test sample data. It should be noted that in the design process of the microwave filter, tuning is usually the most time consuming and may take several months. In contrast, the efficiency of this approach is sufficient even without parallel computing. If parallel computing is applied, the efficiency is quite high. In Figure 7, the comparison between EM responses and 1D-CAE network structure prediction is shown. It can be observed that the two results are completely consistent. Table 1 shows a comparison of the two approaches (EM responses and 1D-CAE network structure prediction) in terms of CPU time. The EM responses time is viewed through the HFSS simulation process information interface after the simulation. 1D-CAE network structure prediction time is calculated by Equation (15). It can be observed that 1D-CAE not only achieves almost the same accuracy as HFSS simulation but also completes faster than EM simulation.

$$time_{1D-CAE} = \frac{time\_total}{n}, \quad (15)$$

where *time_total* is the total predicted time of 1D-CAE. n is the number of 1D-CAE predictions over this total time.

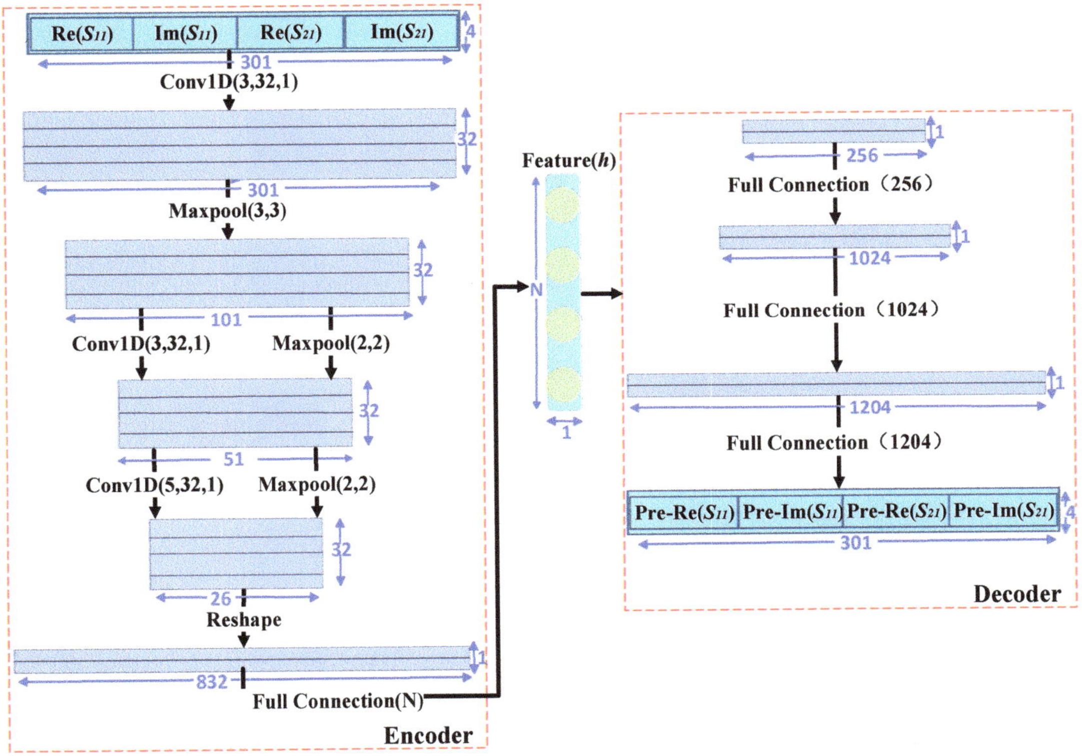

Figure 6. Detailed parameters of 1D−CAE network structure.

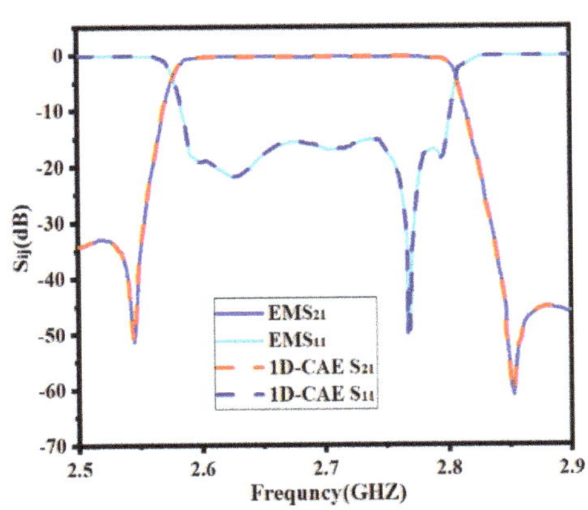

Figure 7. Comparison between full−wave EM simulation result and 1D−CAE prediction result of the first example.

Table 1. Comparison of CPU time between HFSS and 1D-CAE.

	EM Simulation	1D-CAE Prediction
Completion time	3 min	0.017 s

Once the surrogate model is trained well, it can be applied to the MOEA/D, in which the geometric parameters can be adjusted repeatedly. In MOEA/D, multiple design objectives of the filter are set as multiple subproblems. A low-complexity surrogate model based on 1D-CAE is constructed to form the results of the MOEA/D scalar subproblem, which tremendously improves the design efficiency. According to the design requirements of the filter, three objective functions are set for optimization:

$$Fit_1 = \max\{(f_l \leq db(S_{11}) \leq f_h)\} - (-20), \tag{16}$$

$$Fit_2 = \min\{(f_1 - 0.05 \leq db(S_{21}) \leq f_1 + 0.05)\} - (-50), \tag{17}$$

$$Fit_3 = \min\{(f_2 - 0.05 \leq db(S_{21}) \leq f_2 + 0.05)\} - (-50), \tag{18}$$

where f_l is 2.6 GHz and f_h is 2.8 GHz. f_1 = 2.552 GHz and f_2 = 2.858 GHz are the finite transmission zeros of this sixth-order ceramic bandpass filter.

About MOEA/D, the number of neighbors of each weight vector is set to 5, the population size is set to 350, and the maximum number of iterations is set to 60. In order to clearly show the optimization trend, every 10 iterations are defined as a stage, and the optimization results are extracted once. In the sixth stage, MOEA/D found the optimal solution of this sixth-order ceramic bandpass filter. Detailed general flow of this bandpass filter design is shown in Figure 8. The geometric parameters change in the whole process is shown in Table 2. The EM responses at different stages are shown in Figure 9.

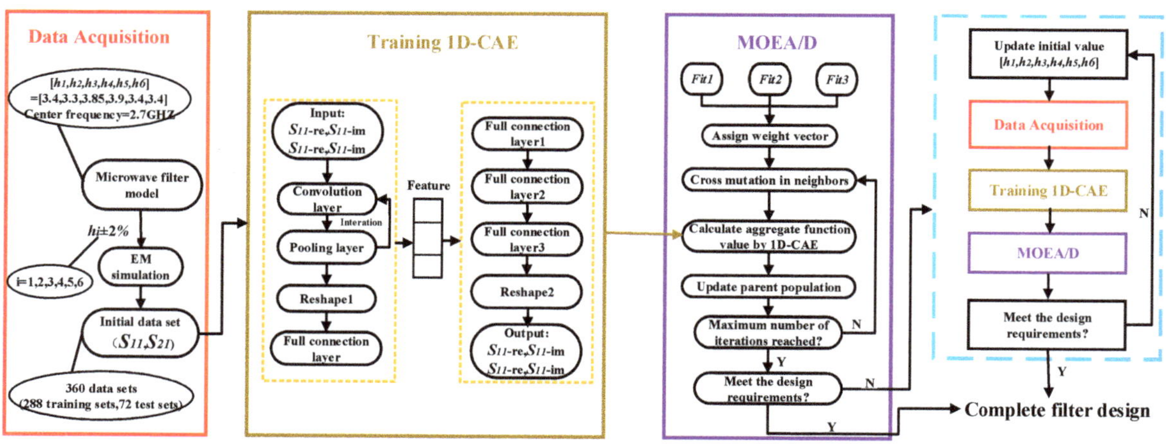

Figure 8. General design flow of the filter.

Table 2. The geometric parameters' change in the whole process.

Stage	*h1*	*h2*	*h3*	*h4*	*h5*	*h6*
0	3.400	3.300	3.850	3.900	3.400	3.400
1	3.374	3.365	3.845	3.883	3.370	3.407
2	3.350	3.370	3.845	3.853	3.370	3.385
3	3.335	3.370	3.845	3.845	3.370	3.360
4	3.330	3.370	3.845	3.845	3.370	3.357
5	3.330	3.369	3.845	3.845	3.370	3.348
6	3.329	3.369	3.845	3.845	3.373	3.330

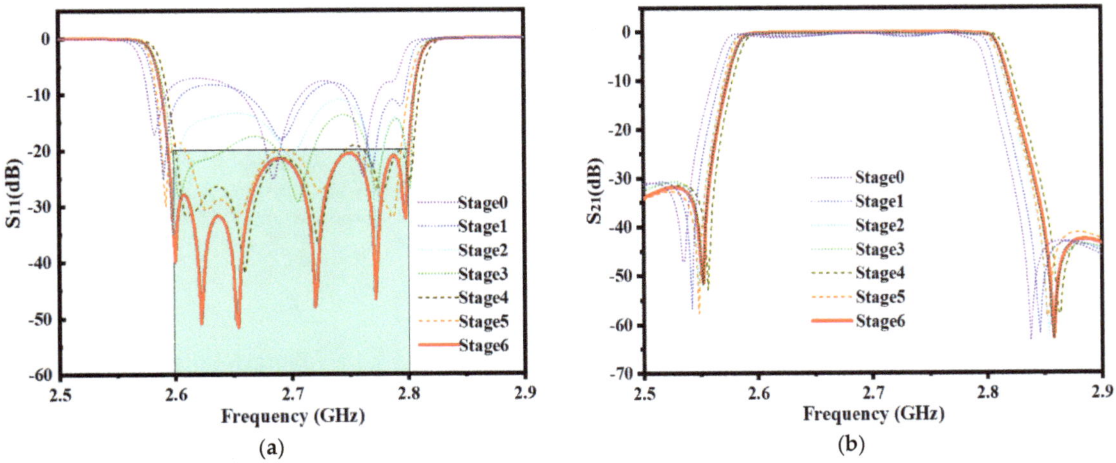

Figure 9. EM responses at different steps of the first example. (**a**) Return loss *S11*. (**b**) Insertion loss *S21*.

The comparisons of the proposed optimization and directly using the EM solver for MOEA/D are shown in Table 3. (The data values in Table 3 are missing because using the EM solver directly does not require training a surrogate model.) We can see that our proposed method can achieve an optimal EM solution in less time.

Table 3. Comparison of CPU time between two optimization design methods.

	Directly Using EM Simulation for MOEA/D	Proposed Optimization
Total EM simulation time	184.5 h	22.5 h
Time of surrogate model training	–	5 min
MOEA/D optimization time	185h	3.5min
Total time	185h	22.64h

3.2. Seventh-Order Metal Cavity Bandpass Filter

In the second example, let us consider a seventh-order metal cavity bandpass filter with a center frequency of 1.791 GHz and a bandwidth of 30 MHz, as shown in Figure 10. The finite transmission zero of this seventh-order metal cavity bandpass filter is f_1 = 1.765 GHz. The design geometric variables are $H = [w_{12}/w_{67}, w_{23}/w_{56}, w_{34}/w_{45}, h_1/h_7, h_2/h_6, h_3/h_5, h_4]^T$. The design requirements of the bandpass filter are as follows:

- $|S11| \leq -20$dB, for 1.776 GHz $\leq \omega \leq$ 1.806 GHz;
- $|S21| \leq -90$dB, for ω = 1.765 GHz;
- $|S21| \leq -40$dB, for ω = 1.825 GHz.

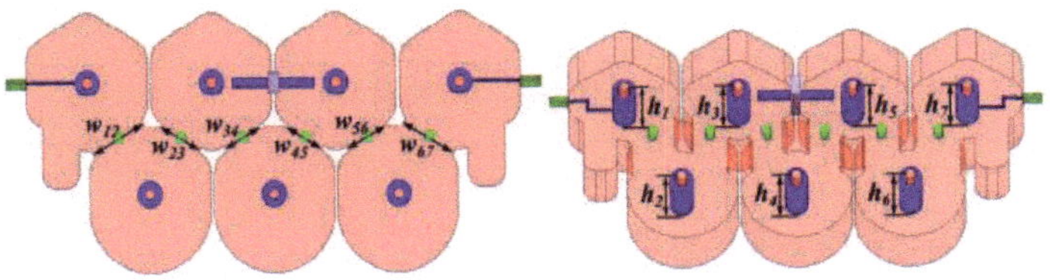

Figure 10. Structure of seventh-order metal cavity bandpass filter.

The starting geometric parameter value of this example is H = [22.2, 17.23, 15.7, 6.63, 5.36, 5.6, 5.7]^T (mm), and its EM response indicates that the response at the starting point is far away from the design requirements.

The parameter values of each layer of 1D-CAE network structure are consistent with Figure 6. So as to guarantee the high accuracy of the 1D-CAE network structure, 550 sets of data, including 440 training sets and 110 test sets, are obtained by HFSS. In Figure 11, the comparison of EM responses with 1D-CAE network structure prediction is shown. Table 4 shows a comparison for the two approaches (EM responses and 1D-CAE network structure prediction) in terms of the CPU time.

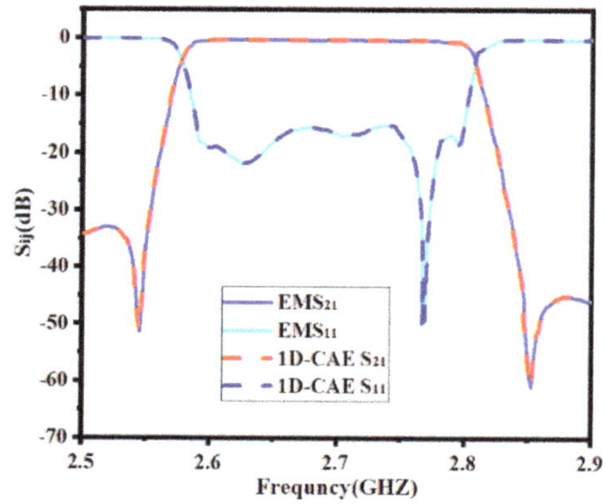

Figure 11. Comparison between full−wave EM simulation result and 1D−CAE prediction result of the second example.

Table 4. Comparison of CPU time between HFSS and 1D-CAE.

	EM Simulation	1D-CAE Prediction
Completion time	12 min	0.02 s

According to the design requirements of this filter, three objective functions are set for optimization:

$$Fit_1 = \max\{(f_l \leq db(S_{11}) \leq f_h)\} - (-20), \tag{19}$$

$$Fit_2 = \min\{(f_1 - 0.05 \leq db(S_{21}) \leq f_1 + 0.05)\} - (-90), \tag{20}$$

$$Fit_3 = \min\{(f_2 - 0.05 \leq db(S_{21}) \leq f_2 + 0.05)\} - (-40), \tag{21}$$

where f_l is 1.776 GHz, f_h is 1.806 GHz, f_1 = 1.765 GHz, and f_2 = 1.825 GHz.

MOEA/D parameters are the same as in the first example, but the maximum number of iterations is set to 50. In order to clearly show the optimization trend, every 10 iterations are defined as a stage, and the optimization results are extracted once. In the fifth stage, MOEA/D found the optimal solution. The geometric parameters' change in the iterative process is shown in Table 5. The EM responses at different stages are shown in Figure 12.

Table 5. The geometric parameters' change in the whole process.

Stage	w_{12}/w_{67}	w_{23}/w_{56}	w_{34}/w_{45}	h_1/h_7	h_2/h_6	h_3/h_5	h_4
0	22.200	17.230	15.700	6.630	5.360	5.600	5.700
1	22.192	17.233	15.679	6.586	5.398	5.657	5.720
2	22.507	17.155	15.684	6.586	5.398	5.660	5.720
3	22.505	17.153	15.689	6.581	5.397	5.643	5.723
4	22.500	17.158	15.690	6.581	5.401	5.645	5.721
5	22.443	17.154	15.689	6.580	5.401	5.645	5.721

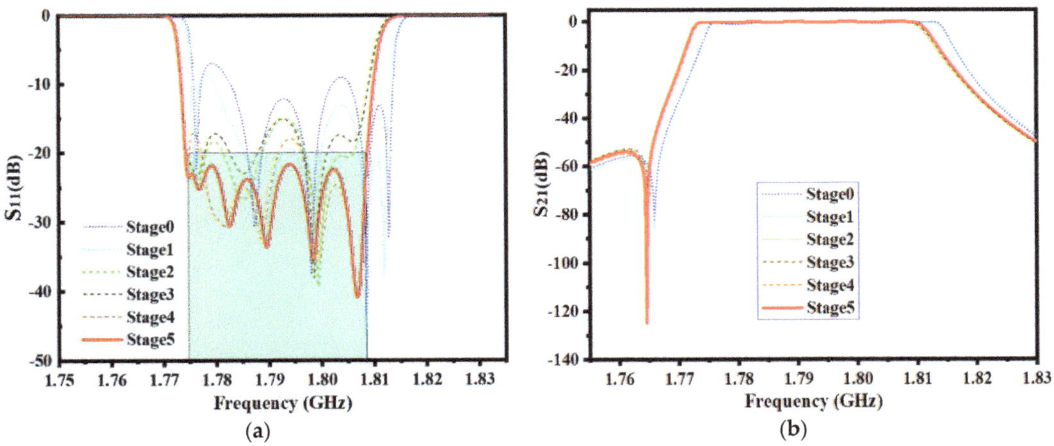

Figure 12. EM responses at different steps of the second example. (**a**) Return loss *S11*. (**b**) Insertion loss *S21*.

The comparisons of the proposed optimization and directly using the EM solver for MOEA/D are shown in Table 6. (The data values in Table 6 are missing because using the EM solver directly does not require training a surrogate model.) We can see that our proposed method can achieve an optimal EM solution in less time.

Table 6. Comparison of CPU time between two optimization design methods.

	Directly Using EM Simulation for MOEA/D	Proposed Optimization
Total EM simulation time	615.5 h	110 h
Time of surrogate model training	–	7 min
MOEA/D optimization time	616 h	5 min
Total time	616 h	110.2 h

4. Conclusions

A novel method merging MOEA/D and a 1D-CAE network structure is proposed to design microwave filters. In this new design method, the 1D-CAE network structure is introduced to replace the traditional full-wave simulation, which greatly reduces the design time of the filter. MOEA/D disassembles the MOPs into a set of single-objective optimization subproblems with neighborhood relations, and each subproblem is a dissimilar set of all objectives. Then, by analyzing the information of adjacent problems, all of the subproblems are optimized by an evolutionary algorithm concurrently. This new method is used to design two microwave filters in our examples. The design results show that the new method merging MOEA/D and the 1D-CAE network structure can be used to design filters that meet the requirements easily. With the help of the 1D-CAE network structure, the MOEA/D algorithm more easily avoids falling into the local optimum and completes the filter design in a shorter time than the EM-based microwave filter design.

In future work, we will continue our research from the following two aspects. First, we will try to apply the proposed technique to the design of other passive devices, such as antennas, couplers, duplexers, etc. Second, we will seek more reliable optimization methods to improve the generalization ability of surrogate models.

Author Contributions: Conceptualization, Y.W. (Yongfeng Wei), Y.Z., G.Q., and Yanxing Wang; methodology, Y.W. (Yongfeng Wei), Y.Z., G.Q., and Y.W. (Yanxing Wang); software, Y.W. (Yongfeng Wei), Y.Z., G.Q., and N.Y.; validation, Y.W. (Yongfeng Wei), Y.Z., G.Q., and N.Y.; formal analysis, Y.W. (Yongfeng Wei), Y.Z., and L.F.; investigation, N.Y.; resources, Y.W. (Yongfeng Wei), Y.Z., G.Q., and Y.W. (Yanxing Wang); data curation, Y.W. (Yongfeng Wei), Y.Z., G.Q.; writing—original draft preparation, Y.W. (Yongfeng Wei), Y.Z., and G.Q.; writing—review and editing, Y.W. (Yongfeng Wei), Y.Z., and L.F.; visualization, Y.W. (Yongfeng Wei), Y.Z., and G.Q.; supervision, Y.W. (Yongfeng Wei) and Y.Z.; project administration, Y.W. (Yongfeng Wei) and Y.Z.; funding acquisition, Y.W. (Yongfeng Wei) and Y.Z. All authors have read and agreed to the published version of the manuscript.

Funding: This work was funded by the National Natural Science Foundation of China (NSFC) under project no. 61761032 and no. 62161032 and the Nature Science Foundation of Inner Mongolia under contract no. 2019MS06006. This work was funded by the Inner Mongolia Foundation 2020MS05059 and Inner Mongolia Department of Transportation NJ-2017-8. This work was also supported by the Shaanxi Key Laboratory of Deep Space Exploration Intelligent Information Technology under grant no. 2021SYS-04. This work was also supported by the Research and Development of New Energy Vehicle Product Testing Conditions in China—Hohhot Baotou, 2017.

Institutional Review Board Statement: Not applicable.

Informed Consent Statement: Not applicable.

Data Availability Statement: Not applicable.

Conflicts of Interest: The authors declare no conflict of interest.

References

1. Cai, Y.X.; He, Y.J.; Zhou, H.W.; Liu, J.J. Design method of LCL filter for grid-connected inverter based on particle swarm optimi-zation and screening method. *IEEE Trans. Power Electron.* **2021**, *36*, 10097–10113. [CrossRef]
2. Zhang, D.H.; You, X.M.; Liu, S.; Yang, K. Multi-colony ant colony optimization based on generalized jaccard similarity recommendation strategy. *IEEE Access* **2019**, *7*, 157303–157317. [CrossRef]
3. Chen, X.; Tian, Y.; Zhang, T.; Gao, J. Differential evolution based manifold gaussian process machine learning for microwave filter's parameter extraction. *IEEE Access* **2020**, *8*, 146450–146462. [CrossRef]
4. Goudos, S.K.; Sahalos, J.N. Pareto optimal microwave filter design using multi-objective differential evolution. *IEEE Trans. Antennas Propag.* **2010**, *58*, 132–144. [CrossRef]
5. Yang, S.Y.; Ho, S.L.; Ni, G.Z.; Machado, J.M.; Wong, K.F. A new implementation of population based incremental learning method for optimizations in electromagnetics. *IEEE Trans. Magn.* **2007**, *43*, 1601–1604. [CrossRef]
6. Ho, S.L.; Yang, S. A population-based incremental learning method for robust optimal solutions. *IEEE Trans. Magn.* **2010**, *46*, 3189–3192. [CrossRef]
7. Yang, L.; Zhu, L.; Choi, W.-W.; Tam, K.-W.; Zhang, R.Q.; Wang, J.P. Wideband balanced-to-unbalanced bandpass filters synthetically designed with Chebyshev filtering response. *IEEE Trans. Microw. Theory Technol.* **2018**, *66*, 4528–4539. [CrossRef]
8. Li, X.; Zhang, X.; Lin, F. Multi-objective design of output LC filter for buck converter via the coevolving-AMOSA algorithm. *IEEE Access* **2020**, *9*, 11884–11894. [CrossRef]
9. Zobaa, A.F. Optimal multiobjective design of hybrid active power filters considering a distorted environment. *IEEE Trans. Ind. Electron.* **2013**, *61*, 107–114. [CrossRef]
10. Chen, B.-S.; Lee, M.-Y.; Chen, X.-H. Security-enhanced filter design for stochastic systems under malicious attack via smoothed signal model and multiobjective estimation method. *IEEE Trans. Signal Process.* **2020**, *68*, 4971–4986. [CrossRef]
11. Zhang, Q.; Li, H. MOEA/D: A multiobjective evolutionary algorithm based on decomposition. *IEEE Trans. Evol. Comput.* **2007**, *11*, 712–731. [CrossRef]
12. Fan, Z.; Li, W.; Cai, X.; Hu, K.; Lin, H.; Li, H. Angle-based constrained dominance principle in MOEA/D for constrained multi-objective optimization problems. *IEEE Congr. Evol. Comput.* **2016**, 460–467. [CrossRef]
13. Qi, Y.; Ma, X.; Liu, F.; Jiao, L.; Sun, J.; Wu, J. MOEA/D with adaptive weight adjustment. *Evol. Comput.* **2014**, *22*, 231–264. [CrossRef]
14. Li, K.; Deb, K.; Zhang, Q.; Kwong, S. An evolutionary many-objective optimization algorithm based on dominance and decomposition. *IEEE Trans. Evol. Comput.* **2014**, *19*, 694–716. [CrossRef]
15. Li, K.; Zhang, Q.; Kwong, S.; Li, M.; Wang, R. Stable matching-based selection in evolutionary multi-objective optimization. *IEEE Trans. Evol. Comput.* **2014**, *18*, 909–923.
16. Feng, F.; Zhang, C.; Gongal-Reddy, V.-M.; Zhang, Q.-J.; Ma, J. Parallel space-mapping approach to EM optimization. *IEEE Trans. Microw. Theory Technol.* **2014**, *62*, 1135–1148. [CrossRef]

17. Feng, F.; Zhang, C.; Zhang, S.; Gongal-Reddy, V.-M.; Zhang, Q.-J. Parallel EM optimization approach to microwave filter design using feature assisted neuro-transfer functions. In Proceedings of the 2016 IEEE MTT-S International Microwave Symposium (IMS), San Francisco, CA, USA, 22–27 May 2016; pp. 1–3. [CrossRef]
18. Zhang, C.; Feng, F.; Gongal-Reddy, V.-M.; Zhang, Q.J.; Bandler, J.W. Cognition-driven formulation of space mapping for equal-ripple optimization of microwave filters. *IEEE Trans. Microw. Theory Technol.* **2015**, *63*, 2154–2165. [CrossRef]
19. Koziel, S.; Ogurtsov, S.; Bandler, J.W.; Cheng, Q. Reliable space-mapping optimization integrated with EM-based adjoint sensitivities. *IEEE Trans. Microw. Theory Technol.* **2013**, *61*, 3493–3502. [CrossRef]
20. Jin, J.; Feng, F.; Na, W.; Zhang, J.N.; Zhang, W.; Zhao, Z.H.; Zhang, Q.-J. Advanced cognition-driven EM optimization incorporating transfer function-based feature surrogate for microwave filters. *IEEE Trans. Microw. Theory Technol.* **2021**, *69*, 15–28. [CrossRef]
21. Zhang, J.; Feng, F.; Na, W.; Jin, J.; Zhang, Q.J. Adaptively weighted training of space-mapping surrogates for accurate yield estimation of microwave components. In Proceedings of the 2020 IEEE/MTT-S International Microwave Symposium (IMS), Los Angeles, CA, USA, 18–27 June 2020; pp. 64–67. [CrossRef]
22. Zhang, W.; Feng, F.; Jin, J.; Zhang, Q.-J. Parallel multiphysics optimization for microwave devices exploiting neural network surrogate. *IEEE Microw. Wirel. Components Lett.* **2021**, *31*, 341–344. [CrossRef]
23. Zhang, C.; Na, W.; Zhang, Q.J.; Bandler, J.W. Fast yield estimation and optimization of microwave filters using a cogni-tion-driven formulation of space mapping. In Proceedings of the 2016 IEEE MTT-S International Microwave Symposium (IMS), San Francisco, CA, USA, 22–27 May 2016; pp. 1–4.
24. Zhang, J.; Feng, F.; Jin, J.; Zhang, Q.-J. Efficient yield estimation of microwave structures using mesh deformation-incorporated space mapping surrogates. *IEEE Microw. Wirel. Components Lett.* **2020**, *30*, 937–940. [CrossRef]
25. Feng, F.; Zhang, J.; Zhang, W.; Zhao, Z.; Jin, J.; Zhang, Q.-J. Coarse-and fine-mesh space mapping for EM optimization incorporating mesh deformation. *IEEE Microw. Wireless Compon. Lett.* **2019**, *29*, 510–512. [CrossRef]
26. Sans, M.; Selga, J.; Rodríguez, A.; Bonache, J.; Boria, V.E.; Martín, F. Design of planar wideband bandpass filters from specifications using a two-step aggressive space mapping (ASM) optimization algorithm. *IEEE Trans. Microw. Theory Technol.* **2014**, *62*, 3341–3350. [CrossRef]
27. Wang, Y.; Zhang, Z.; Yi, Y.; Zhang, Y. Accurate microwave filter design based on particle swarm optimization and one-dimensional convolution autoencoders. *Int. J. RF Microw. Comput. Eng.* **2021**, *32*, e23034. [CrossRef]
28. Zhang, Z.; Liu, B.; Yu, Y.; Cheng, Q.S. A microwave filter yield optimization method based on off-line surrogate model-assisted evolutionary algorithm. *IEEE Trans. Microw. Theory Technol.* **2022**, *70*, 1–10. [CrossRef]
29. Masci, J.; Meier, U.; Cireşan, D.; Schmidhuber, J. Stacked convolutional auto-encoders for hierarchical feature extraction. In *Lecture Notes in Computer Science*; Springer: Berlin/Heidelberg, Germany, 2011; pp. 52–59.
30. Geng, J.; Fan, J.C.; Wang, H.Y.; Ma, X.R.; Li, B.M.; Chen, F.L. High-resolution SAR image classification via deep convolu-tional autoencoders. *IEEE Trans. Geosci. Remote Sens.* **2015**, *12*, 2351–2355. [CrossRef]
31. Park, S.; Yu, S.; Kim, M.; Park, K.; Paik, J. Dual autoencoder network for retinex-based low-light image enhancement. *IEEE Access* **2018**, *6*, 22084–22093. [CrossRef]
32. Zhao, L.J.; Huang, H.K.; Li, X.; Ding, S.X.; Zhao, H.L.; Han, Z.Y. An accurate and robust approach of device-free local-ization with convolutional autoencoder. *IEEE Internet Things J.* **2019**, *6*, 5825–5840. [CrossRef]
33. Yu, J.; Zhou, X. One-dimensional residual convolutional autoencoder based feature learning for gearbox fault diagnosis. *IEEE Trans. Ind. Informatics* **2020**, *16*, 6347–6358. [CrossRef]
34. Wen, T.; Zhang, Z. Deep convolution neural network and autoencoders-based unsupervised feature learning of EEG signals. *IEEE Access* **2018**, *6*, 25399–25410. [CrossRef]

electronics

MDPI

Article

Research on Multi-Strategy Routing Protocol in Flying Ad Hoc Networks

Ning Xin [1,*], Xiaoqun Chen [1], Te Chen [1], Feng Liu [1], Kun Liu [1] and Yuxiong Lin [2]

1 Institute of Telecommunication and Navigation Satellites, China Academy of Space Technology, Beijing 100094, China
2 School of Computer Science and Engineering, Northeastern University, Shenyang 110169, China
* Correspondence: xinning7@sina.com; Tel.: +86-010-68112474

Abstract: In this paper, a multi-strategy routing protocol, reactive-greedy-face (RGF), is proposed based on the advantages of reactive forwarding, greedy forwarding, and projected face forwarding strategies. This protocol improves and unites the strategies through a reasonable selection and switching mechanism to make up for the deficiencies of the existing three-dimensional routing protocols in the flight self-assembly network and improve the rapid recovery of packet forwarding after topology changes. The simulation validation shows that this routing protocol can be used to recover packets after a topology change. Simulation verification indicates that the routing protocol significantly improves the packet delivery rate and the average end-to-end delay performance, and can better adapt to the flight self-assembly network's three-dimensional high-dynamic and low-density characteristics.

Keywords: unmanned aerial vehicle; flying ad hoc networks; routing protocol; three-dimensional high-dynamic network; reactive mechanism

Citation: Xin, N.; Chen, X.; Chen, T.; Liu, F.; Liu, K.; Lin, Y. Research on Multi-Strategy Routing Protocol in Flying Ad Hoc Networks. *Electronics* **2022**, *11*, 3327. https://doi.org/10.3390/electronics11203327

Academic Editor: Dimitris Kanellopoulos

Received: 6 September 2022
Accepted: 13 October 2022
Published: 15 October 2022

1. Introduction

Recently, the unmanned aerial system (UAS) has set off a development upsurge in the world; it has a wide range of applications in photography, entertainment, emergency rescue, counter-terrorism, stability maintenance, and agricultural remote sensing [1–4]. With the increasing demand for UAVs in the civil and military fields, the UAS uses many multi-UAV cooperation methods and has furthered cluster, information, and intellectual development. As with a single UAV system, multi-UAV systems can also allow UAVs to communicate by connecting with a ground control station (GCS) or satellite [5]. However, the hardware equipment needed by UAVs to communicate through satellites is expensive, and there are limits to the range of communication through ground control stations. To solve the problem of communication between multiple UAVs, an effective solution is to introduce the technology of ad hoc networks. The flying ad hoc network (FANET) [6] proposed by İlkerBekmezci et al. is a fast and efficient solution for realizing the networking of multiple UAVs by using ad hoc technology. In the FANET, some UAVs can communicate with a ground control station, and the other part can communicate with a satellite. All UAVs eventually form a self-organizing network. This way, UAVs can communicate with each other and with ground control stations. The FANET is a three-dimensional high-dynamic network. In existing mobile ad hoc networks and wireless sensor networks (WSNs), the research on two-dimensional geographic location routing protocols has been quite mature. Therefore, many of the three-dimensional geographic location routing protocols in the FANET are the expansion of corresponding two-dimensional protocols in a three-dimensional space, but the effect of expansion is limited. For example, greedy routing in a three-dimensional topology is more likely to fall into local minima than greedy routing in a two-dimensional topology, that is, to encounter a routing hole, which leads to greedy routing failure. Durocher et al. proved that there is no deterministic local memoryless routing algorithm in a three-dimensional space that can ensure the delivery of data packets [7], which also means that the uncertain recovery routing algorithm can ensure that data packets jump out of the routing hole. Therefore, the design of an efficient three-dimensional geographic location routing protocol is still facing great challenges.

Many researchers have studied the operation of traditional ad hoc network routing protocols in the FANET and have achieved some results, but they still need to improve and develop new routing protocols according to the characteristics and requirements of the FANET. In [8], the traditional ad hoc on-demand distance vector routing (AODV) protocol can be improved by adding the location and heading information of the source node as well as a minimum expected connection time (minCT) and maximum risk value to the RREQ packet as a joint measure to select the best route. The reactive green reactive (RGR) protocol proposed in [9] combines a reactive routing mechanism with greedy geographic forwarding. Once the reactive path is interrupted, the existing packets will be saved by greedy geographic forwarding during the re-routing. The RGR protocol effectively reduces the delay while maintaining the overhead of the conventional routing level. In the literature [10–12], the AntHocNet protocol, based on the ant colony algorithm, and the BeeAdHoc protocol, based on the bee colony algorithm, were analyzed. Simulation results showed that the heuristic algorithm based on swarm intelligence is more suitable for the flight ad hoc network characterized by high-speed data transmission and high node mobility. The authors of [13] proposed a predictive routing algorithm for the flight ad hoc network based on the clustering structure. The algorithm uses the improved Dijkstra algorithm to calculate the minimum end-to-end delay path, which avoids link loss in the transmission process. In [14], an adaptive beacon scheme called ABPP (adaptive beacon and position prediction) was proposed for the geographic location routing protocol. ABPP uses historical location information to predict the future location of UAVs through a weighted linear regression model, which effectively reduces the channel overhead. In [15], a greedy location-assisted routing scheme based on fountain code was proposed, which was called FGPA (fountain code-based green position assisted routing). FGPA takes the relative motion trend between two UAV nodes as a new measure. The authors of [16] proposed an adaptive forwarding protocol (AFP) in a 3D space. AFP can reduce unnecessary replay and conflict between neighbor nodes based on forwarding probability and region standards. In [17], the hybrid routing protocol HSRP (hierarchical space routing protocol) was proposed based on the ZRP protocol. The HSRP protocol uses the flow rate mechanism to select table-driven or on-demand routing protocols. In [18], aiming at the topological structure in the dynamic change of an ad hoc network, the topological network structure of the recording platform was designed by using the random constant speed moving model (RCS), which made the simulation scene closer to the dynamic performance of the high-dynamic mobile node. In [19], a novel routing protocol inspired by the cuckoo search algorithm was proposed. The latest meta-heuristic cuckoo search algorithm (CSA) was used to improve the AODV routing protocol and identify the shortest path between two nodes. The authors of [20] proposed a FANET routing protocol, PSO-GLFR, based on greedy forwarding and limited flooding. Particle swarm optimization was used to solve the problem of greedy forwarding suboptimal selection and realize low delay and low power routing support. In [21], a geographic location-based routing protocol was proposed to reduce routing overhead using neighbor geographic location information. In [22], a multi-data rate mobility aware (MDRMA) protocol was proposed. The MDRMA protocol is the primary extension of a self-organizing distance vector in two-way mobile sensing with adaptive HELLO message protocol, which can quickly generate stable routes and reduce link failures. In [23], an adaptive greeting interval scheme (EE-HELLO) was proposed, reducing energy consumption under the same packet transmission rate. In [24], based on dynamic source routing (DSR), a continuous Hopfield neural network was used to optimize routing to adapt to the FANET network. The optimized DSR protocol improved the key performance of routing. In [25], an adaptive hybrid communication protocol was proposed, including position prediction-based directional MAC (PPMAC) and a self-learning method. The protocol provides a highly autonomous communication solution.

Most three-dimensional geographic location routing protocols use two working modes simultaneously: the greedy mode (greedy routing) and the recovery mode (recovery routing, also known as bypass mode or bypass routing). The greedy mode ensures packets are forwarded in advance, while the recovery mode is mainly used to make packets jump out of routing holes or loops when the greedy mode encounters a local minimum or routing loops. The three-dimensional geographic location routing protocol will generally follow the following basic principles: Each node knows its 3D location information. In some cases, nodes can use a virtual location instead of an actual location to simplify the configuration. Currently, the sending node knows the location information of the surrounding neighbor node and the destination node. Each node needs to exchange location information with the neighbor node, which can be achieved by periodically sending beacon messages. The current routing decision of the sending node is based on local information and is fully distributed. Generally, the current sending node selects the next-hop forwarding node according to its own location, the neighbor node's location, and the destination node's location. The current sending node uses the greedy mode to decide on the next-hop forwarding node. The common way this is achieved is through greedy geographic forwarding (GGF) [26]. For the switch of the recovery mode, when the current sending node encounters a local minimum or routing loop, the greedy mode will not continue to work. At this time, the protocol needs to switch to the recovery mode to forward packets until it jumps out of the routing hole or loop and then switches back to greedy mode. According to the above principles, the three-dimensional geographical location routing protocol can be designed reasonably, which can efficiently complete the communication task of the flight ad hoc network. However, the structure of the 3D dynamic network is complex. Making 3D geographic location routing work efficiently with a reasonable switch routing strategy has become a hot and arduous task in the FANET.

However, previous FANET routing protocols have focused on selecting the best path through joint metrics, although they use greedy and recovery modes. Although some researchers have considered network topology dynamics, most researchers have determined routing paths with the idea of studying the UAV location or trajectory prediction, for example. With the rapid changes in the routing environment brought about by the highly dynamic motion of UAVs, few researchers have conducted studies. At this stage, there is a lack of routing protocols for fast strain in different topological situations caused by highly dynamic flying UAVs. We have built on previous research to design routing switching strategies to adapt to highly dynamic changing UAV networks based on different network scenarios brought about by highly dynamic UAV flights. The previous study also improved the routing strategy to make it more suitable for the FANET.

According to the characteristics of a three-dimensional network, when routing protocols encounter different network environments, routing forwarding can be effectively completed by selecting different routing strategies. Compared with the previous scheme, the proposed three-dimensional geographic location routing protocol, RGF, can complete the communication task of the FANET more effectively. The overall idea of the protocol is to provide a low overhead and scalable location services through the routing discovery process of the HELLO packet active flooding strategy and reactive fisheye technology, and to effectively improve the location acquisition ability of the 3D network. The reactive path in the 3D network is easily interrupted. At this time, the improved greedy forwarding can immediately transmit data packets and improve the link duration of the 3D network. However, greedy forwarding easily falls into local minima and forms routing holes in three-dimensional networks. Multi-path projection face forwarding can help packets jump out of routing holes. Projection face can easily generate path expansions, which can avoid potential path expansions by switching to reactive mechanism forwarding. The simulation results show that RGF can effectively improve the routing and forwarding capability of the FANET.

The main contributions of this paper are summarized as follows:

1. We propose a multi-strategy routing RGF protocol to solve the topological high-dynamic problem of the FANET by reasonably selecting and switching different forwarding policies.
2. We propose a hybrid location service strategy that can achieve the low overhead, scalable location services required by the RGF protocol.
3. We improve the traditional reactive path-building mechanism and greedy geographic forwarding mechanism to make the selected path or next-hop node more reasonable.
4. We propose a multi-path projected face forwarding policy and directionally adaptive forwarding rules to avoid routing loops, reduce path sprawl, and improve the efficiency of jumping out of routing holes.

The research scheme is shown in Figure 1. The rest of this paper is arranged as follows: Section 2 introduces the relevant research content, Section 3 is the location service design of the routing protocol, Section 4 is the forwarding strategy selection and handover method, Section 5 is the specific forwarding strategy design, and Section 6 is the simulation analysis of the experiment.

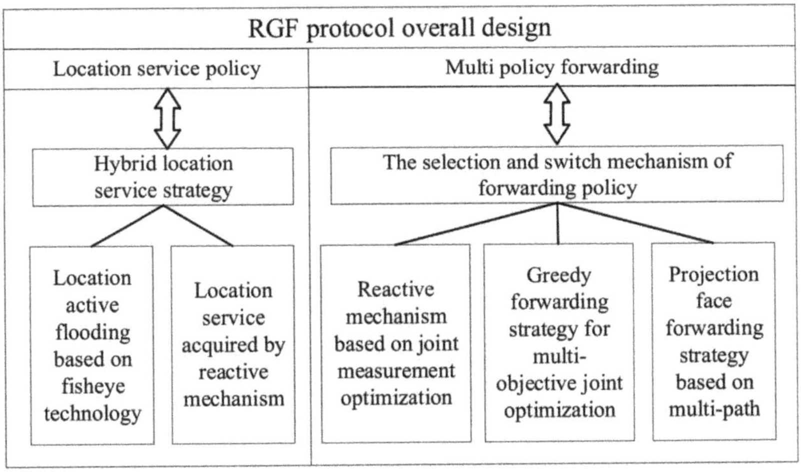

Figure 1. RGF protocol overall design.

2. Hybrid Location Service Strategy Design

Although the flooding-based method is practical in some traditional location-based service methods, it needs a high consumption of network resources. It is difficult to apply the set-based method to high-dynamic networks because the nodes in the update set and search set will constantly change, so it is necessary to constantly adjust the node set at the intersection. However, the method based on attribution generally needs some fixed nodes or infrastructure as location servers, which is not conducive to the deployment and expansion of the network.

Therefore, a hybrid location service strategy was used in the RGF protocol, which combined proactive and reactive mechanisms. In a certain range, the location information and motion information between nodes are exchanged in the way of active flooding. Outside the range, the target node's location information and motion information are obtained on demand by building a reactive path, which effectively controls the cost of flooding and ensures the scalability of this location-based service scheme.

In the proactive location service, an active flooding strategy based on fisheye technology was adopted, as shown in Figure 2.

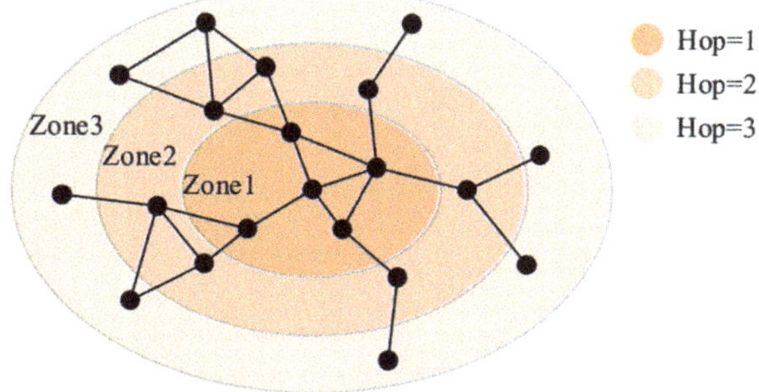

Figure 2. Scope of fisheye.

Each node needs periodic flooding HELLO messages, and the remaining forwarding times field controls the size of the flooding area. Each time the intermediate node forwards a HELLO message, it is necessary to reduce the value of the remaining forwarding times by 1. When the value of the remaining forwarding times is 0, the HELLO message will not be forwarded. Therefore, the coverage area of HELLO messages can be controlled by setting the initial number of the remaining forwarding times, reducing the cost and ensuring the scalability of the proactive location service. The average update frequency in different areas will be different. The closer the node gets, the higher the average update frequency, and the more accurately the location and motion information can be maintained.

In addition, considering the high mobility of UAVs, the faster the UAV moves, the more frequently the position changes. Therefore, to obtain the latest location information in time, the update cycle of the HELLO message should be short. The node in AODV is usually set to route outage when it does not receive HELLO messages from the next-hop node twice in a row. The HELLO message interval is usually set to 1 s. Therefore, if the next-hop node has moved out of the communication range of the current node, the current node will only know after 2 s, so the packets transmitted during this temporary blind period will be lost. Therefore, the update period of the HELLO message should be less than or equal to 0.5 s. Here, Equation (1) is used to adjust the HELLO message's update cycle dynamically.

$$T_{HELLO} = T_{MIN} + \beta^{\alpha^{|\vec{v}|}}, \tag{1}$$

In the formula, T_{HELLO} represents the current update cycle of the HELLO message, T_{MIN} is the minimum update cycle of the HELLO message, and $|\vec{v}|$ represents the UAV's motion rate, $0 < \beta < 1$, $\alpha > 1$.

When the destination node is out of the fisheye range, it will not be able to obtain the HELLO message of the destination node regularly, and the source node will not be able to obtain the location information and motion information of the destination node. Therefore, we need to obtain the information of the destination node through the reactive location service. The reactive location service process is shown in Figure 3, which is realized through the route discovery process of the reactive mechanism.

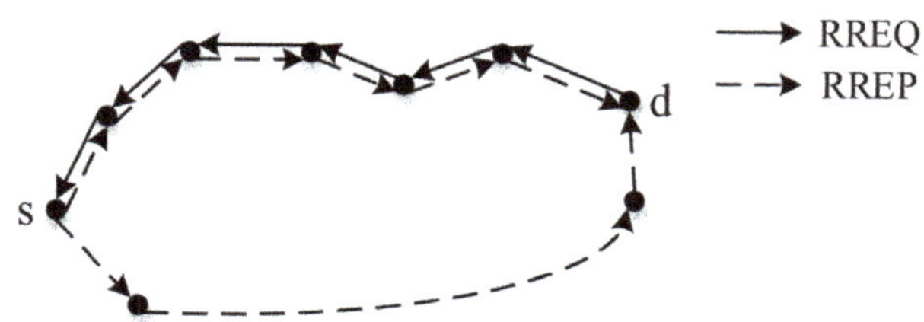

Figure 3. Reactive location service.

In reactive location service, the source node and the destination node use the reactive mechanism's RREQ message and RREP message for route discovery to build a reactive path. Among them, the RREQ message encapsulates the location information, motion information, time stamp, and effective time of the source node, and the RREP message encapsulates the location information, motion information, time stamp, and effective time of the destination node. Therefore, through the interaction of the two messages, the source node will obtain the location and motion information of the destination node. The reactive path constructed will also be used for the subsequent packet forwarding of the reactive mechanism.

3. Forwarding Policy Selection and Handover Design

In the RGF protocol, each packet contains a packet header to control the packet forwarding policy. The main fields of the packet header are shown in Figure 4. The source node needs to determine the initial forwarding policy for these packets, and the intermediate node must also change the forwarding policy according to the situation.

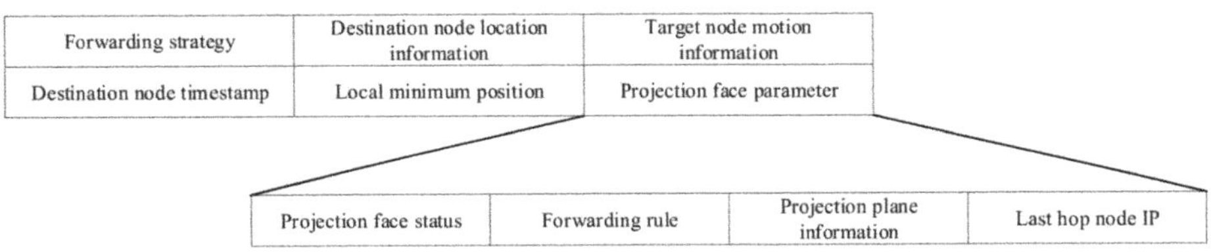

Figure 4. RGF data packet header format.

3.1. Forwarding Policy Selection

In the RGF protocol, the source node must select and set the initial forwarding policy of the packet, as shown in Figure 5.

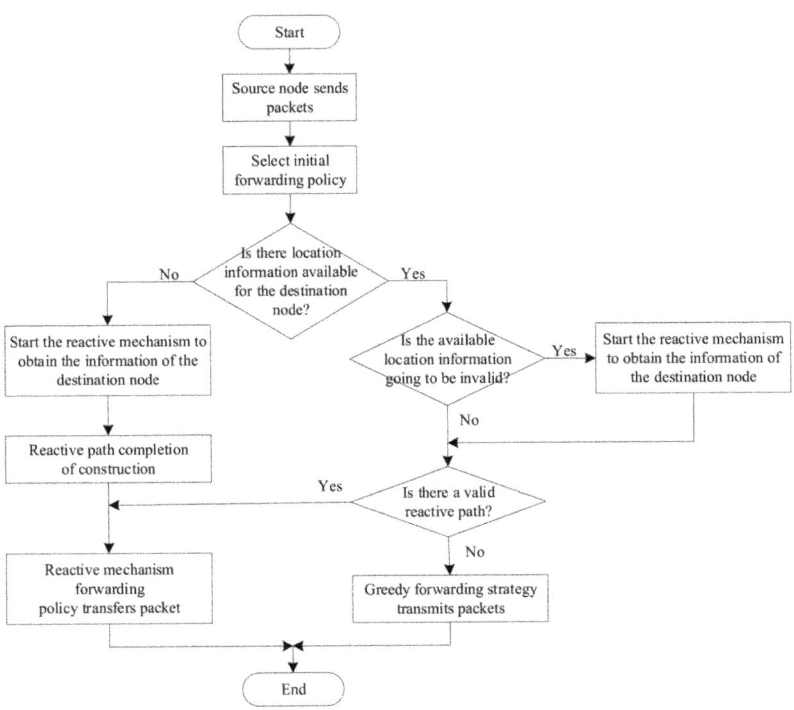

Figure 5. The process by which the source node selects the forwarding strategy.

Suppose the source node does not have the effective location information of the destination node. In that case, it needs to start the reactive mechanism of the hybrid location service to build a reactive path and obtain the location information of the destination node, and then use the path to execute the reactive mechanism to forward the policy to transmit the data packet. If the source node has valid location information for the destination node, and there is also a reactive path available, the path will also be used first to transmit packets. Suppose the source node has valid location information for the destination node, but there is no available transmission path. In that case, the greedy forwarding strategy will be used to transmit the packets directly. In addition, if the source node finds that the location information of the destination node is about to be invalid, it actively obtains the location and motion information of the destination node through the reactive location service. At the same time, the current packet can still encapsulate the effective information of the destination node for transmission, thus avoiding the waiting delay caused by the need to retrieve the information of the destination node.

The process of selecting the forwarding policy for intermediate nodes is shown in Figure 6.

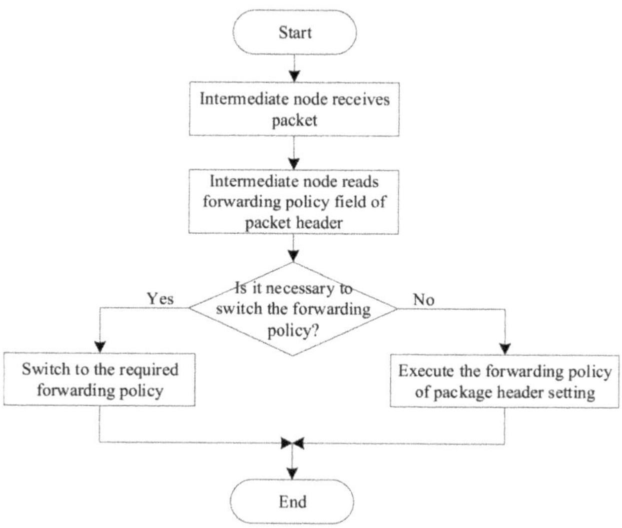

Figure 6. The process by which the intermediate node selects the forwarding strategy.

When an intermediate node forwards a packet, it first selects the forwarding policy of the packet according to the forwarding policy field in the packet header. Only when the handover conditions are met can the intermediate node change the forwarding policy of the packet. Otherwise, even if the intermediate node has a reactive path to the destination node, if the packet is currently greedy forwarding and the intermediate node is not the minimum local node, the intermediate node also needs to forward the packet with a greedy forwarding strategy.

3.2. *Forwarding Policy Switching*

When the current packet forwarding strategy is the reactive mechanism, the intermediate node must check whether the next-hop node on the reactive path is still in the communication range. If it is, the intermediate node will continue to perform reactive mechanism forwarding and forward the packet to the corresponding next-hop node. Otherwise, it will switch to greedy forwarding, as shown in Figure 7.

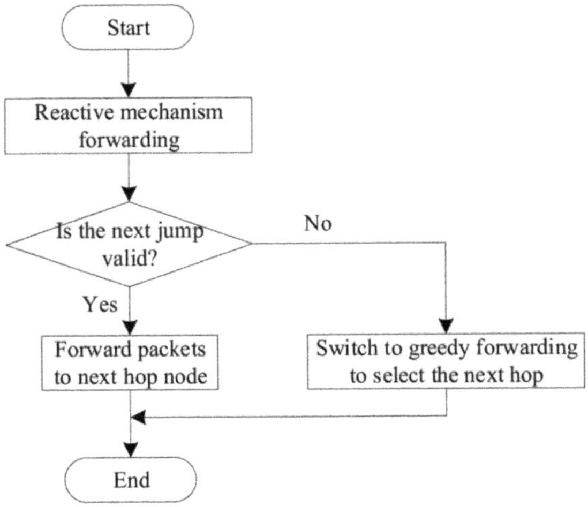

Figure 7. Switching conditions under the reactive mechanism.

When the current packet forwarding strategy is greedy forwarding, the intermediate node must determine whether it is a minimum local node. If it is not, it will continue to use greedy forwarding. If it is, it will check whether there is a reactive path available. If there is, it will switch to reactive mechanism forwarding. Otherwise, it will use projection face forwarding, as shown in Figure 8.

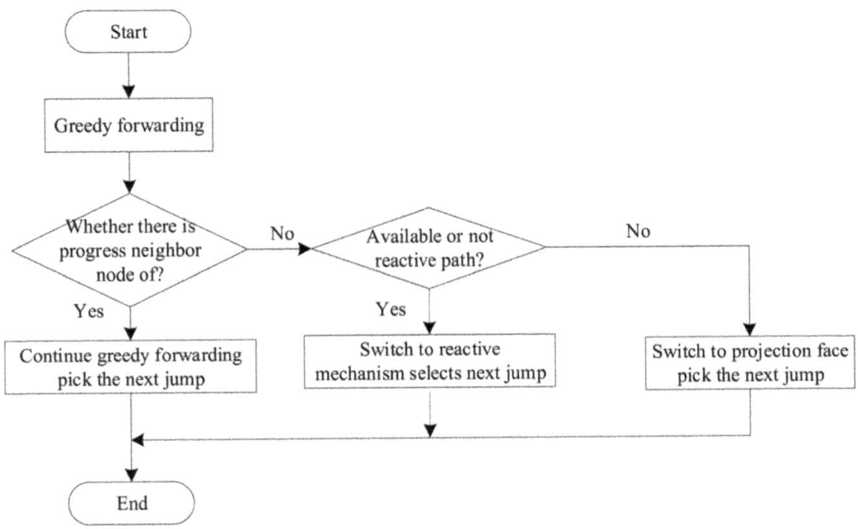

Figure 8. Switching conditions under greedy forwarding.

When the current packet forwarding strategy is to project face, the intermediate node first needs to determine whether it has jumped out of the routing hole. If so, the intermediate node will switch back to greedy forwarding. If not, then it will determine whether there is a reactive path available. If so, it will switch to reactive mechanism forwarding. If not, it will continue to perform projected face forwarding, as shown in Figure 9.

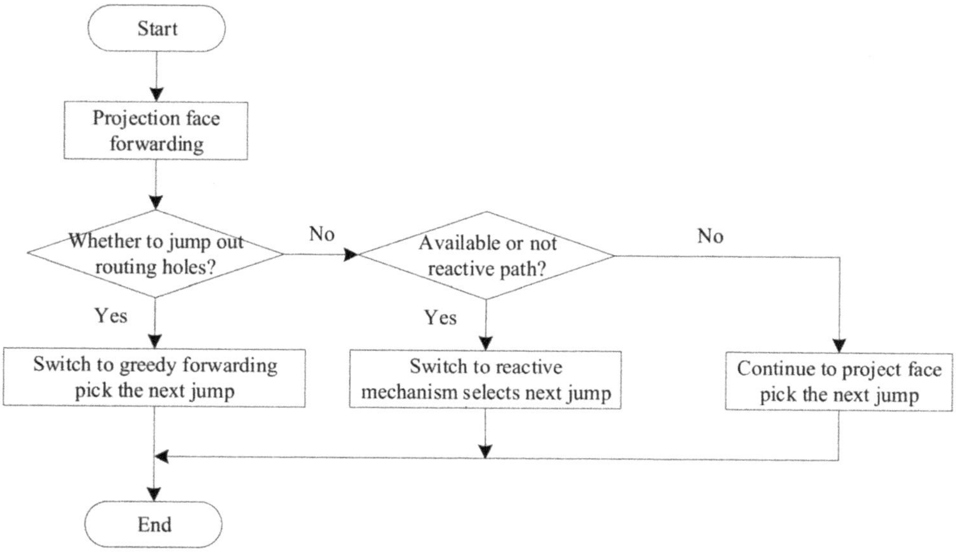

Figure 9. Switching conditions under the projection face.

4. Forwarding Strategy Design

4.1. Reactive Mechanism Design Based on Joint Measurement Optimization

Due to the highly dynamic nature of the FANET, the network topology can change at any time. Therefore, to ensure the stability of the constructed reactive path, the RGF protocol needs to add a minimum link expiration time word segment in the RREQ message and RREP message of the reactive mechanism used. If node i and node j are in communication range of each other, the three-dimensional position and speed vectors of the nodes are (x_i, y_i, z_i) and (v_{ix}, v_{iy}, v_{iz}), respectively. The three-dimensional position and velocity vectors of node j are (x_j, y_j, z_j) and (v_{jx}, v_{jy}, v_{jz}), respectively. Then, the link expiration time (LET) between node i and node j is calculated according to Equations (2) and (3).

$$LET_{ij} = \frac{\sqrt{(d^2+e^2+f^2)R^2 - (ae-bd)^2 - (af-cd)^2 - (bf-ce)^2} - (ad+be+cf)}{d^2+e^2+f^2} \quad (2)$$

$$\begin{cases} a = x_i - x_j \\ b = y_i - y_j \\ c = z_i - z_j \\ d = v_{ix} - v_{jx} \\ e = v_{iy} - v_{jy} \\ f = v_{iz} - v_{jz} \end{cases}, \quad (3)$$

When forwarding RREQ or RREP, each intermediate node needs to calculate the link expiration time with the previous hop node and then compare it with the value in the minimum link expiration time field. If it is less than the value recorded in this field, it will be replaced with a new value. Therefore, the "minimum link expiration time" field in RREQ or RREP will record the minimum value of all link expiration times on the path that the message passes, that is, the expiration time of the path, where RREQ records the expiration time on the reverse path to the source node, and RREP records the expiration time on the forward path to the destination node.

The destination node must start a short timer after receiving the first RREQ. According to the received RREQ during the timing period, a joint metric is calculated for each path according to Equation (4). After the timer expires, the path with the largest joint metric value is activated by RREP.

$$pathT = \frac{mLET}{n}, \tag{4}$$

In Equation (4), $pathT$ represents the joint metric, $mLET$ represents the minimum link expiration time, and n represents the number of path hops. The path selected according to Equation (4) will have a certain stability, which can avoid the handoff of the forwarding strategy as much as possible, and can consider the number of hops, which can ensure a lower delay.

After receiving the RREQ or RREP, other intermediate nodes also need to calculate the joint metric value for the newly discovered forward or reverse path according to Equation (4). Suppose there is active forward or reverse routing information in the routing table. In that case, intermediate nodes also need to calculate the current joint metric value on the old path as recorded in Equation (5).

$$pathT_{current} = pathT_{record} - \frac{T_{current} - T_{record}}{n}, \tag{5}$$

In the formula, $pathT_{current}$ is the latest joint measure of the old path, $pathT_{record}$ is the joint measure recorded in the routing table when the old path is constructed, $T_{current}$ is the current time, T_{record} is the time stamp recorded in the routing table, and K is the number of hops recorded in the routing table.

When one of the following conditions is met, the intermediate node needs to update the routing table:

a. The new serial number is greater than the destination node's serial number stored in the routing table;

b. The two serial numbers are equal, but the joint measure of the new path is greater than the current joint measure of the old path;

c. The new serial number is an unknown serial number.

The design algorithm of the reaction mechanism is as follows:

Algorithm 1: The path construction process of the reactive mechanism of the RGF protocol.

1. Source node or minimum local node sends RREQ, where the minimum link expiration time field is set to a negative value, indicating infinity.
2. After receiving an RREQ, the intermediate node first adds 1 to the hop value of the RREQ, then calculates the link expiration time with the previous hop node, and finally compares it with the value in the minimum link expiration time field in the RREQ. If it is less than the minimum link expiration time in the RREQ, then the intermediate node alternates the values. Otherwise, it does nothing.
3. The intermediate node calculates the joint metrics of the new and old paths and updates the routing table's request node entries according to the routing table's update conditions. For RREQs that have been forwarded, if the recorded path is better, it is allowed to forward the RREQ repeatedly.
4. After receiving the first RREQ, the destination node starts a timer, during which the joint measurement value of the corresponding path is calculated for each received RREQ.
5. After the timer expires, the destination node selects the path with the largest joint metric value and sends an RREP for activation. The minimum link expiration time field is also set to a negative value.
6. After receiving the RREP, the intermediate node also judges whether it is necessary to update the minimum link expiration time field of the RREP. At the same time, it updates the destination node entry of the routing table according to the update conditions of the routing table and then looks up the request node entry to forward the RREP to the previous hop node.
7. When the source node or the minimum local node receives the RREP of the destination node, it transmits the packet with the activated reactive path.

Algorithm 2: RGF protocol reactive mechanism forwarding packet flow

1. The intermediate node checks whether it is the packet's destination node. If it is, it receives the packet. Otherwise, it continues.
2. The intermediate node checks and updates the destination node information in the packet header or its routing table.
3. The intermediate node queries the destination node entry of the routing table, finds the next-hop node on the reactive path, and then calculates the current location of the next-hop node according to the relevant information in the node information table.
4. The packet will be forwarded if the next-hop node is in communication range. Otherwise, the intermediate node will modify the forwarding policy field in the packet header and switch to greedy forwarding.
5. The above steps are repeated until the packet is delivered to the destination node.

4.2. Greedy Forwarding Strategy Design Based on Multi-Objective Joint Optimization

RGF encapsulates the destination node's location and motion information in the packet header, so it can switch to a greedy forwarding strategy when the reactive path is interrupted. The greedy forwarding of the RGF protocol adopts the greedy geographical forwarding method, and the forwarding node needs to forward the packets to the neighbor node closer to the destination node than itself. In order to ensure the stability of packet transmission, when the RGF protocol performs greedy forwarding, the forwarding node also needs to calculate the corresponding link expiration time for each neighbor node closer to the destination node according to Equations (2) and (3), to select the next hop with a stable link.

In addition, if the sending queue of the next-hop node is long, the packet will be sent to the node, which will obviously increase its load. At the same time, if the remaining energy of the next-hop node is low, the packets will be forwarded to the node, which will cause the node to die faster, and the packets that the node has not yet sent will also be lost. Therefore, considering load balancing and energy saving, each node's HELLO message also encapsulates the sending queue length and remaining energy of the node. The forwarding node will finally calculate a transmission cost for each neighbor node closer to the destination node according to Equation (6) and select the node with the lowest cost as the next hop.

$$cost = \rho_1 \frac{D}{D_f} + \rho_2 L + \frac{\rho_3}{LET} + \frac{\rho_4}{E}, \tag{6}$$

In the formula, *cost* represents the transmission cost, D represents the distance from the neighbor node to the destination node, D_f represents the distance from the current node to the destination node, L represents the normalized transmission queue length of the neighbor node, LET represents the link expiration time between the current node and the neighbor node, E represents the remaining energy of the neighbor node, and ρ_1,ρ_2,ρ_3,ρ_4 is the corresponding weight and meets the requirements of $0 < \rho_1, \rho_2, \rho_3, \rho_4 < 1$ and $\rho_1 + \rho_2 + \rho_3 + \rho_4 = 1$.

According to Equation (6), there will be a relatively stable link between the next-hop node selected by greedy forwarding and the current node. The node's load is low, and the remaining energy is also relatively sufficient, so it can ensure the reliability of packet transmission, reduce the delay of intermediate forwarding, and eliminate the nodes with low power consumption to achieve a multi-objective joint optimization. The specific algorithm is as follows:

Algorithm 3: RGF protocol greedy forwarding packet process

1. The intermediate node checks whether it is the packet's destination node. If it is, it receives the packet. Otherwise, it continues.
2. The intermediate node checks and updates the destination node information in the packet header or its routing table.
3. According to the information in the node information table, the intermediate node calculates the current location of all hop neighbor nodes. It calculates the current location of the destination node according to the destination node information in the packet header.
4. The intermediate node calculates the distance between each hop neighbor node and the destination node and the distance between itself and the destination node, and finds a hop neighbor node closer to the destination node to form a set, which is called the forward node set, and the nodes in it are also called forward nodes.
5. If the forward node set is empty, the current node is the minimum local node. The current node checks whether there is an effective reactive path. If there is one, the current node will modify the forwarding policy field in the packet header to forward the packet with a reactive mechanism; otherwise, it will forward the packet with a projected face and send the RREQ to the destination node simultaneously to build an available reactive path through route discovery.
6. If the forward node set is not empty, the transmission cost of each forward node is calculated according to Equation (6), and the node with the lowest cost is selected as the next hop.
7. The above steps are repeated until the packet is delivered to the destination node.

4.3. Design of Projection Face Forwarding Strategy Based on Multi-Path

When greedy forwarding encounters a local minimum, and no available reactive path needs to be re-routed, the RGF protocol will use projection face as a temporary forwarding strategy to transmit the received packets during reactive path establishment. Because projection face forwarding only needs to use the information within the local scope, it will not bring additional overhead, and it is relatively flat or has node density in the network. When it is high, projection face forwarding can show a better performance. However, in the FANET, because of the strong mobility of nodes, the rapid change in topology, and the low density of nodes, the direct use of traditional projection face forwarding based on a single path will still bring higher path expansion and increase the delay of packet delivery. Therefore, the RGF protocol improves the traditional projection face forwarding in three aspects to improve its performance.

4.3.1. Selection of Projection Nodes

In the projection face forwarding of the RGF protocol, only the 3DRNG or 3DGG neighbor node of the current forwarding node is used for projection. That is to say, the network topology is simplified by constructing a 3DRNG graph or 3DGG graph. Then, the one-hop neighbor node after topology simplification is projected vertically onto the projection plane so that a local clear projection face can be obtained after projection, and there is no cross. Since the next-hop forwarding node must be selected from the 3DRNG or 3DGG neighbor nodes, the forwarding rules can still be executed normally.

4.3.2. Construction and Forwarding of Multi-Projection Plane

To improve the projection face forwarding out of the routing hole, the projection face forwarding of the RGF protocol adopts a multi-path strategy: constructing multiple projection planes simultaneously and transmitting a corresponding copy of the data packet on different projection planes. As shown in Figure 10, the minimum local node needs to encapsulate its current location in the local minimum location field of the packet header and then broadcast the packet to all its 3DRNG or 3DGG neighbor nodes. These nodes construct the corresponding projection plane according to their position, the position of the destination node, and the position of the local minimum value recorded in the packet header, save the information of the plane in the projection face parameter field of the packet header, and then forward the received copy of the packet on the projection plane.

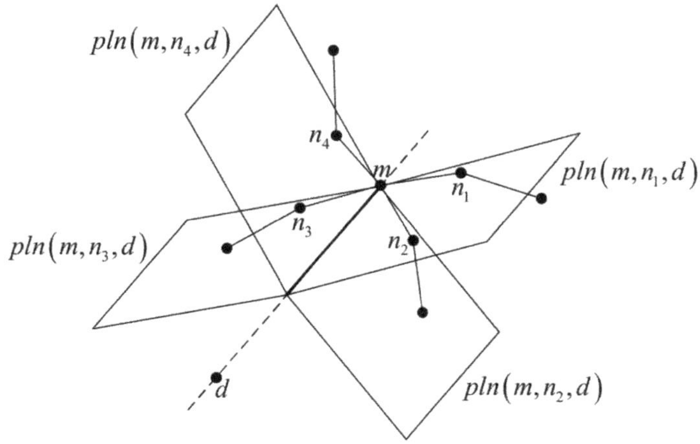

Figure 10. Construction of multi-projection plane.

Because of the projection of multiple planes, and the same group of nodes in different projection planes, the next-hop node selected may be different, so the occurrence of the loop can be avoided to some extent. Because multiple copies of packets are transmitted at the same time, the probability of packets jumping out of the routing hole can be increased, and the path expansion can be reduced.

4.3.3. Direction-Adaptive Forwarding Rules

The projection face of the RGF protocol adopts a rule called direction-adaptive forwarding (DAF). The DAF rule enables the node n_3 in Figure 11 to rotate the link with the node m in a clockwise direction so that the forwarding node of the next hop is n_4, while the node selection n_1 remains unchanged.

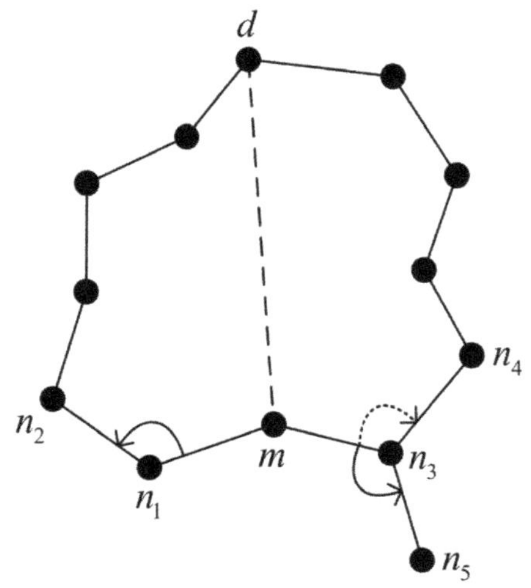

Figure 11. Wrong search area.

The DAF rule combines the right-hand rule and left-hand rule. The left-hand rule is shown in Figure 12. The current node finds the next-hop node in the plan by rotating the link with the previous hop node clockwise around itself, so it can generally traverse the interior of a face anticlockwise. X, Y and Z represent different nodes.

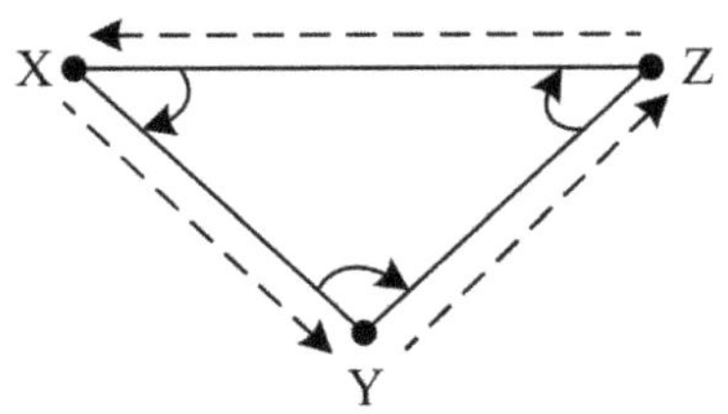

Figure 12. Left-hand rule.

The DAF rule is composed of two sub-rules. Sub-rule 1 determines whether the initial forwarding rule is a right-hand rule or a left-hand rule, while sub-rule 2 determines when it is necessary to switch the forwarding rule from the right-hand rule/left-hand rule to the left-hand rule/right-hand rule. Suppose on the projection plane, the location of the local minimum value recorded in the packet header is m, the projection of the destination node is d, the projection of a 3DRNG or 3DGG neighbor node of the local minimum value node is n, the connection between point m and node d is md, the connection between point m and node n is mn, a node in the forwarding process is r, the last-hop node of the node r is s, and the connection between point r and node s is rs. Then, the two sub-rules of the DAF rule are described as follows:

Sub-rule 1: if the connecting line md rotates counterclockwise around the touchpoint m (the angle should be less than 180°) to the connecting line mn, then the node n and its subsequent nodes will traverse the face with the right-hand rule; if the connecting line md rotates clockwise around the touchpoint m (the angle should be less than 180°) to the connecting line mn, then the node n and its subsequent nodes will traverse the face with the left-hand rule.

Sub-rule 2: if node s forwards the packets to node r through the right-hand rule/left-hand rule, and node r detects that there is a cross between line rs and line md, then node r and the nodes after it need to switch to the left-hand rule/right-hand rule.

Figure 13 shows an example of the DAF rule, in which node n_1 and node n_2 decide the initial forwarding rule according to sub-rule 1, and node r switches the right-hand rule to the left-hand rule according to sub-rule 2 and implements face switching.

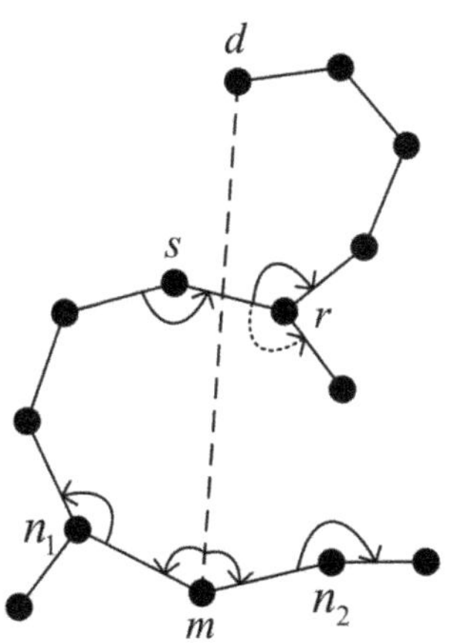

Figure 13. An example of DAF rules.

DAF rules expand the search area of projection face forwarding, reduce the possibility of routing loops and repeated searches, improve the probability of packets jumping out of routing holes, and also, to a certain extent, inhibit path expansion and reduce the delivery delay of packets. The specific algorithm is as follows:

Algorithm 4: RGF initialization process of projection face forwarding protocol

1. The minimum local node encapsulates its current location information into the local minimum location field of the packet header. It sets the projection face state of the projection face parameter field to the construction state and the IP of the previous hop node to its IP, and then broadcasts the packet to the next-hop neighbor node.
2. The one-hop neighbor node of the minimum local node first constructs the 3DRNG graph according to the relevant information. It then determines whether it is the 3DRNG neighbor of the minimum local node. If so, it will continue with the following steps; otherwise, it will discard the data packet.
3. If the 3DRNG neighbor node of the minimum local node judges whether it can switch to greedy forwarding or reactive mechanism forwarding, it stops the initialization of the projection face and switches to the corresponding forwarding strategy. Otherwise, it will continue with the following steps.
4. The 3DRNG neighbor node of the minimum local node sets the projection face state of the projection face parameter field of the packet header to the forwarding state.
5. The 3DRNG neighbor node of the minimum local node constructs the projection plane according to its position, the position of the destination node, and the local minimum position recorded in the packet header and encapsulates the information of the projection plane into the projection plane information of the projection face parameter field of the packet header.
6. The 3DRNG neighbor node of the minimum local node sets the last-hop node's IP in the projection face parameter field as the IP of this node.
7. The 3DRNG neighbor node of the minimum local node determines the initial forwarding rule according to sub-rule 1 of the DAF rule. It saves it to the forwarding rule in the projection face parameter field of the packet header.
8. The local minimum node's 3DRNG neighbor node projects its 3DRNG neighbor node vertically onto the projection plane.
9. The local minimum node's 3DRNG neighbor node selects the next-hop node according to the forwarding rules used and forwards the packets.

Algorithm 5: The projection face of the RGF protocol forwards packets

1. The intermediate node checks whether it is the packet's destination node. If it is, it receives the packet. Otherwise, it continues.
2. The intermediate node checks and updates the destination node information in the packet header or its routing table.
3. The intermediate node calculates the current positions of all its hop neighbor nodes and the previous hop nodes according to the relevant information in the node information table. It calculates the current positions of the destination nodes according to the destination node information in the packet header.
4. If the intermediate node judges whether it can switch to greedy forwarding or reactive mechanism forwarding, it will modify the forwarding policy field in the packet header to switch to the corresponding forwarding policy. Otherwise, it will continue to follow the steps below.
5. The middle node constructs a 3DRNG graph to determine its 3DRNG neighbor nodes.
6. The intermediate node saves its IP to the previous node's IP in the projection face parameter field of the packet header.
7. The middle node reads the projection plane information from the projection face parameter field of the packet header. It then vertically projects this node, the 3DRNG neighbor node, the previous hop node, and the destination node onto the projection plane.
8. The intermediate node judges whether the connection between itself and the previous hop node crosses the connection between the local minimum value position and the destination node on the projection plane. If yes, it switches the forwarding rules according to sub-rule 2 of the DAF rule and saves the switched forwarding rules to the forwarding rules in the projection face parameter field of the packet header. Otherwise, it will continue to perform the original transfer rules.
9. The intermediate node selects the next-hop node according to the forwarding rules and forwards the packets.
10. The above steps are repeated until the packet is delivered to the destination node.

5. Experimental Results and Discussion

In this experiment, OMNeT++ software was used to build the scene. UAV nodes were randomly distributed, and different random topologies were generated by setting different random seeds. The basic parameter settings are shown in Table 1.

Table 1. Simulation parameters.

Parameters	Value
Simulation time	Total 1000 s (10 topologies, each run for 100 s)
Scene size	10,000 m × 10,000 m × 6000 m
Packet size	1024 Bytes
Packet generation rate	8–16/s
Node number	35, 40, 45, 50, 55, and 60
Communication radius	2000 m
Transmission speed	2 Mbps
Mobility model	RWP
Moving speed	50–100 m/s, 100–150 m/s, 120–150 m/s, 150–200 m/s, 200–250 m/s, 250–300 m/s, and 300–350 m/s
HELLO message interval of RGF	$T_{MIN} = 0.5$, $\alpha = 1.0035$, $\beta = 0.8$
Greedy forwarding parameters of RGF	$\rho_1 = 0.5$, $\rho_2 = 0.1$, $\rho_3 = 0.3$, $\rho_4 = 0.1$

In the existing research on the FANET, the reactive routing protocol AODV, based on the network topology and the greedy geographic forwarding routing protocol (greedy) based on the whole network flood model location service, is widely used. However, the service performance of the above protocols is limited in the three-dimensional high-dynamic network topology environment. The existing common improvements include the combination of intelligent optimization algorithms to improve the routing protocol, such as the AntHocNet [13] routing protocol, and some combined routing protocols, such as the combination of the optimization and greedy PSO-GLFR [21] routing protocol. AntHocNet and GLFR represent the two improved algorithms in the comparison test diagram. This experiment verified the advantages of the proposed RGF routing protocol compared with the traditional routing protocol and the improved routing protocol in the above environment through the simulation of the flight ad hoc network scene by the above parameters. The experiment used three commonly used indicators to analyze the protocol's performance, including the delivery rate, the average delay, and the routing cost, and verified the superiority of each indicator proposed in this paper under different mobile speeds and different nodes.

Figures 14–16 show the comparison experiments of each index of the five protocols under different node-mobility speeds.

Figure 14. The effect of velocity on the delivery rate.

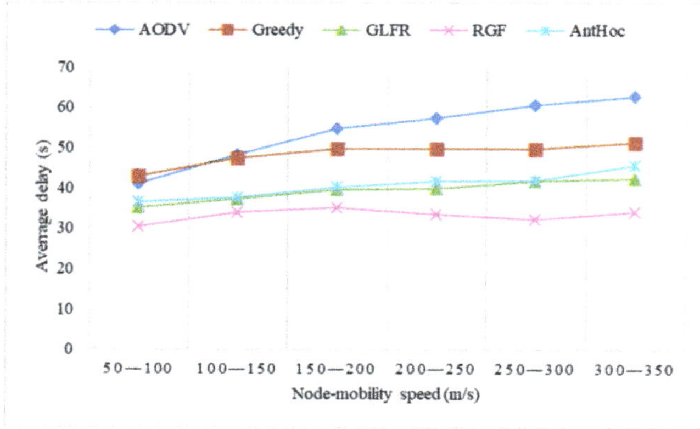

Figure 15. The effect of velocity on average delay.

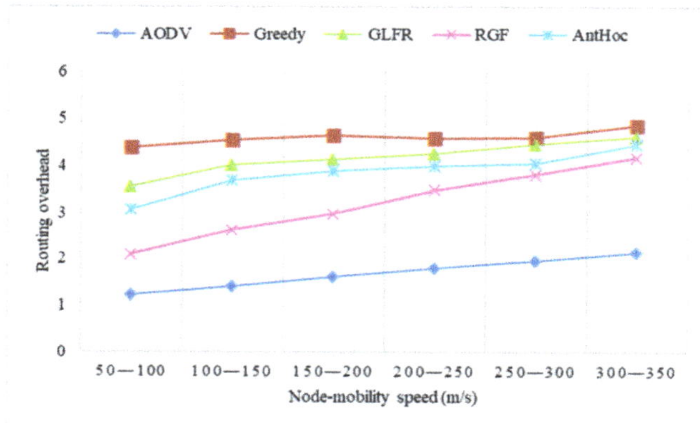

Figure 16. The effect of velocity on routing overhead.

From Figure 14, it can be seen that the delivery rate of all protocols was not high, mainly due to the small number of nodes and the sparsity of the network. As the node movement speed increased, the delivery rate of AODV underwent a significant decrease, which was due to the lack of location awareness in AODV. The faster the nodes move, the more serious the packet loss is. In contrast, greedy, RGF, GLFR, and AntHocNet encapsulated node movement information in HELLO messages so that the current position of the next-hop node could be calculated before transmitting packets. The low delivery rate of greedy was due to the fact that greedy geographic forwarding is more likely to encounter local minima in a 3D scenario when the node can only wait for the routing hole to eliminate itself before continuing to transmit packets. In contrast, RGF can use the projection face forwarding strategy to continue transmitting packets when greedy forwarding encounters a local minimum, and the minimum local node will also start the reactive path construction at the same time and later transmit packets to jump out of the routing hole through the reactive mechanism forwarding strategy. From the figure, we can see that the delivery rate was the lowest when the nodes moved at 150–200 m/s, which was because the packet delivery efficiency was high when the nodes moved at a slower speed. However, after the node movement speed reached a certain level, the node movement speed was not the primary influence in ensuring the delivery rate. Due to faster path changes, routing holes became the main effect. At 150–300 m/s, as the movement speed became faster, path changes helped to break the routing hole. However, after more than 300 m/s, the node movement speed gradually increased in its impact on packet delivery because the location-aware efficiency of the routing protocols gradually decreased after the node speed increased to a certain level, which affected the accurate construction of paths.

Figure 15 demonstrates that with an increase in the node movement speed, the average delay of AODV experienced a significant increase, mainly because the increase in the node movement speed would have led to more frequent changes in the network topology, and route interruptions would have been more likely to occur. Therefore, nodes would need to continuously send RREQs and RREPs to establish new transmission paths, which would cause a larger delay; at the same time, the delivery rate was the worst. GLFR and AntHocNet were not sensitive to topology changes, and the average delay increased more gently, while when the forwarding encountered a local minimum, the node could only passively wait for the recovery of the routing hole, so the average delay was higher than that of RGF. RGF was more stable than the other routing protocols because of the introduction of joint metric optimization in the process of reactive path building, allowing it to reduce the occurrence of route disruptions. Even if the reactive mechanism forwarding encountered route interruption, it could quickly switch to greedy forwarding, and when greedy forwarding encountered a local minimum, it could continue transmitting packets directly with projected face forwarding without additional waiting time, and the minimum local node would send RREQs at this time without causing packet waiting, so the average delay of RGF was lower than that of the other routing protocols. From Figure 15, we can see that the average delay was the highest when the node moved at 150–200 m/s. The reason is the same as the effect of node speed on the delivery rate, where location awareness was the main influence on packet delivery when the node speed was low or high, and the main impact on packet delivery at moderate speeds was jumping out of routing holes.

As can be seen in Figure 16, the routing overhead of AODV, RGF, GLFR, and AntHocNet grew significantly as the node movement speed increased, while the routing overhead of greedy increased only slightly. The delivery rate of AODV decreased rapidly as the node movement speed increased, and because of the frequent route disruptions, a large number of control messages had to be generated to notify the source node as well as re-route discovery, resulting in an increase in the routing overhead, but since flooded HELLO messages were not required, the routing overhead was still the smallest among the five protocols. In greedy, the nodes required periodic network-wide flooding of HELLO messages and thus, it incurred the largest routing overhead, but since the number of flooded HELLO messages was only related to the number of nodes, and the delivery rate was relatively flat, the routing overhead hardly increased. In RGF, the nodes also needed to flood HELLO messages within a certain range, and because of the location-aware capability, they could detect topology changes earlier than AODV, so they also sent more other control packets; therefore, the routing overhead was still higher than that of AODV, despite the higher delivery rate than AODV.

For the experimental results, with an increase in the node-mobility speed, the delivery rate of AODV, greedy, PSO-GLFR, and AntHocNet decreased significantly, while that of RGF was relatively stable. In terms of the average delay, AODV had significant growth, while greedy, PSO-GLFR, AntHocNet, and RGF had a little growth, among which RGF's growth was the smallest. In terms of the re-routing cost, RGF's cost was higher than that of AODV, and greedy's cost was the highest but basically stable.

Figures 17–19 show the comparison experiments of each index of the five protocols under different node numbers.

Figure 17 demonstrates that as the number of nodes in the network increased, the weaker the network sparsity became; therefore, the delivery rate for all protocols improved significantly. For ADOV, the increase in the number of nodes made the network less prone to segmentation and so improved the delivery rate, but the lack of location awareness and the high dynamics of the topology still left a significant number of packets unnecessarily lost during transmission, so the improvement to AODV was the smallest of the four protocols. For the greedy, GLFR, and AntHocNet routing protocols, the increase in the number of nodes reduced the probability of routing holes in the network, so that forwarding could be executed more smoothly, and the delivery rate was higher than that of AODV. RGF relied more on greedy forwarding to transmit packets in highly dynamic networks, so the improvement was similar to that of greedy. Because it could rely on projection face forwarding to jump out of routing holes, the final delivery rate was higher.

As can be seen in Figure 18, the average delay of all protocols decreased significantly with an increase in the number of nodes, which is related to an increase in the delivery rate of these protocols. For AODV and greedy, the increase in the number of nodes allowed the existence of multiple paths to the destination node in the network, and even if the current route was broken, a new transmission path could be found, reducing the delay waiting in between due to the retransmission of control messages. For GLFR and AntHocNet, the increase in the number of nodes made packet forwarding less prone to encountering local minima, and route holes existed for shorter periods of time, thus reducing the waiting time for hole recovery and decreasing the latency. The increase in RGF with the number of nodes allowed the reactive mechanism to have the opportunity to establish more stable transmission paths, and greedy forwarding could be performed more smoothly.

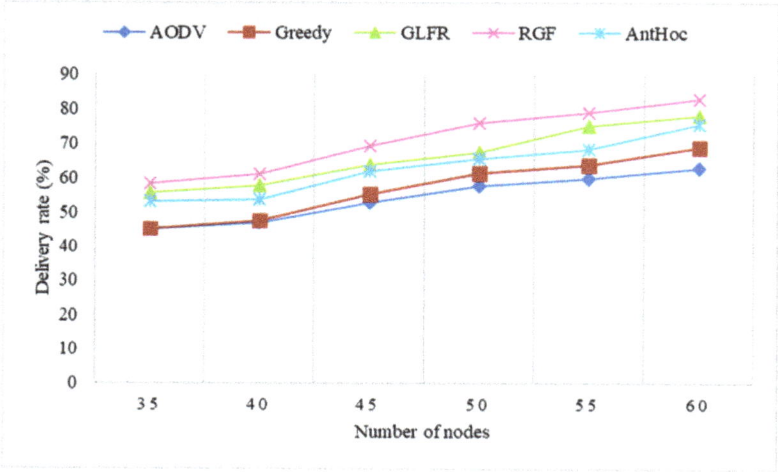

Figure 17. The effect of the number of nodes on the delivery rate.

Figure 18. The effect of the number of nodes on the average delay.

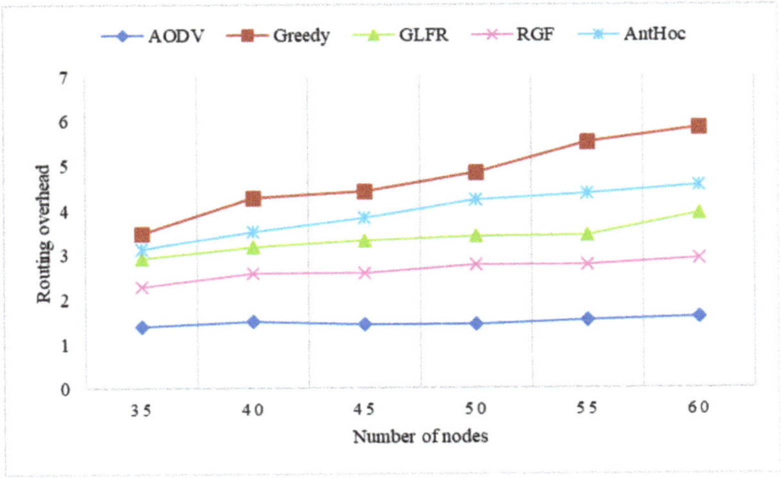

Figure 19. The effect of the number of nodes on routing overhead.

The variation in the routing overhead with the number of nodes is shown in Figure 19. With an increase in the number of nodes, the routing overhead grew rapidly for greedy, with almost no growth for AODV and only a small increase for RGF, GLFR, and AntHocNet. For AODV, since route discovery required the network-wide flooding of RREQs, the increase in the number of nodes led to an increase in the number of flooded RREQ messages in the network, but because the delivery rate also increased, the routing overhead remained essentially the same. For greedy, it was clear that the number of successfully received packets did not keep up with the number of control packets as the number of nodes increased due to the need to flood HELLO messages across the network, so the routing overhead increased the most. In contrast, GLFR, RGF, and AntHocNet did not require network-wide flooded HELLO messages, so the growth in the routing overhead was lower than that of greedy.

The experimental results for node number changes are basically close to the results for the node speed changes. From the experimental results, it can be seen that RGF has obvious advantages over AODV, greedy, PSO-GLFR, and AntHocNet in terms of delivery rate, average end-to-end delay, and routing overhead in a three-dimensional high-dynamic topology, at a low density, and when there is an uneven distribution of nodes in the flight ad hoc network. In the high-speed and low-density scenarios, the RGF routing performance improved the communication ability of the FANET.

6. Conclusions

In this paper, the RGF routing protocol was proposed. Because of the high-dynamic topology and low density of nodes, the existing routing protocols could not provide a good performance. In order to make up for the shortcomings of a single forwarding strategy, multi-strategy forwarding can effectively reduce the packet waiting delay caused by the need for re-routing. Among them, reactive mechanism forwarding is provided by a reactive mechanism based on the network topology, such as AODV. The reactive mechanism is not only used for packet forwarding, but also for reactive location services, and it improves the efficiency of protocol operations. On the other hand, because the location service of active flooding provides the location information of neighbor nodes, when the reactive path is interrupted and the forwarding of the reactive mechanism cannot continue, it can quickly switch to greedy forwarding based on geographical location, which ensures the continuous transmission of data packets and avoids the impact of path interruption. When greedy forwarding encounters a local minimum problem, it can switch to projection face forwarding for recovery. In order to reduce the path expansion problem caused by projection face forwarding, a minimum local node will also start reactive mechanism construction, which can jump out of the routing hole to switch to reactive mechanism forwarding.

Author Contributions: Conceptualization, N.X. and X.C.; methodology, T.C., N.X., and Y.L.; software, F.L. and K.L.; validation, X.C., T.C., and F.L.; formal analysis, F.L.; investigation, N.X.; resources, X.C.; data curation, K.L.; writing—original draft preparation, N.X.; writing—review and editing, Y.L.; visualization, T.C.; supervision, Y.L.; project administration, X.C.; funding acquisition, N.X. All authors have read and agreed to the published version of the manuscript.

Funding: This research was funded by the National Key Research and Development Program of China, grant number 2019YFB1803100.

Informed Consent Statement: Not applicable.

Data Availability Statement: The data presented in this study are available on request from the corresponding author.

Conflicts of Interest: The authors declare no conflict of interest.

References

1. Tang, Y.; Miao, Y.; Barnawi, A.; Alzahrani, B.; Alotaibi, R.; Hwang, K. A joint global and local path planning optimization for UAV task scheduling towards crowd air monitoring. *Comput. Netw.* **2021**, *193*, 107913. [CrossRef]
2. Chang, B.; Tang, W.; Yan, X.; Tong, X.; Chen, Z. Integrated Scheduling of Sensing, Communication, and Control for mmWave/THz Communications in Cellular Connected UAV Networks. *IEEE J. Sel. Areas Commun.* **2022**, *40*, 2103–2113. [CrossRef]
3. Xie, Z.; Song, X.; Cao, J.; Qiu, W. Providing Aerial MEC Service in Areas Without Infrastructure: A Tethered-UAV-Based Energy-Efficient Task Scheduling Framework. *IEEE Internet Things J.* **2022**. [CrossRef]
4. Sajid, M.; Mittal, H.; Pare, S.; Prasad, M. Routing and scheduling optimization for UAV assisted delivery system: A hybrid approach. *Appl. Soft Comput.* **2022**, *126*, 109225. [CrossRef]
5. Frew, E.W.; Brown, T.X. Networking Issues for Small Unmanned Aircraft Systems. *J. Intell. Robot. Syst.* **2009**, *54*, 21–37. [CrossRef]
6. Bekmezci, İ.; Sahingoz, O.K.; Temel, Ş. Flying Ad-Hoc Networks (FANETs): A survey. *Ad Hoc Netw.* **2013**, *11*, 1254–1270. [CrossRef]
7. Durocher, S.; Kirkpatrick, D.; Narayanan, L. On routing with guaranteed delivery in three-dimensional ad hoc wireless networks. *Wirel. Netw.* **2010**, *16*, 227–235. [CrossRef]
8. Gankhuyag, G.; Shrestha, A.P.; Yoo, S.J. A novel directional routing scheme for flying ad-hoc networks. In Proceedings of the International Conference on Information and Communication Technology Convergence, Jeju Island, Korea, 19–21 October 2016.
9. Shirani, R.; St-Hilaire, M.; Kunz, T.; Zhou, Y. Combined Reactive-Geographic routing for Unmanned Aeronautical Ad-hoc Networks. In Proceedings of the Wireless Communications and Mobile Computing Conference, Shanghai, China, 27–31 August 2012.
10. Maistrenko, V.A.; Alexey, L.V.; Danil, V.A. Experimental estimate of using the ant colony optimization algorithm to solve the routing problem in FANET. In Proceedings of the International Siberian Conference on Control and Communications, Moscow, Russia, 12–14 May 2016.
11. Leonov, A.V. Application of bee colony algorithm for FANET routing. In Proceedings of the 2016 17th International Conference of Young Specialists on Micro/Nanotechnologies and Electron Devices, Erlagol, Altai, Russia, 30 June–4 July 2016.
12. Leonov, A.V. Modeling of bio-inspired algorithms AntHocNet and BeeAdHoc for Flying Ad Hoc Networks (FANETS). In Proceedings of the International Scientific-Technical Conference on Actual Problems of Electronics Instrument Engineering, Novosibirsk, Russia, 3–6 October 2016.
13. Rovira-Sugranes, A.; Razi, A. Predictive routing for dynamic UAV networks. In Proceedings of the IEEE International Conference on Wireless for Space and Extreme Environments, Montréal, QC, Canada, 10–12 October 2017.
14. Li, X.; Huang, J. ABPP: An Adaptive Beacon Scheme for Geographic Routing in FANET. In Proceedings of the International Conference on Parallel and Distributed Computing, Applications and Technologies, Taipei, Taiwan, 18–20 December 2017.
15. Zhu, Y.; Huang, Q.; Li, J.; Wu, D.O. Design and evaluation of airborne communication networks. In Proceedings of the Seventh International Conference on Ubiquitous and Future Networks, Sapporo, Japan, 7–10 July 2015.
16. Qingwen, W.; Gang, L.; Zhi, L.; Qian, Q. An adaptive forwarding protocol for three dimensional Flying Ad Hoc Networks. In Proceedings of the International Conference on Electronics Information and Emergency Communication, Beijing, China, 14–16 May 2015.
17. Zhong, D.; Wang, Y.; Zhu, Y.; You, T. An Aeronautical Ad Hoc Network routing protocol based on air vehicles movement features. In Proceedings of the International Conference on Applied Electromagnetics and Communications, Dubrovnik, Croatia, 19–21 September 2016.
18. Chen, L.; Ruijuan, Y.; Meirong, H. Ad hoc high-dynamic routing protocol simulation and research. *Wirel. Commun. Netw. Appl.* **2016**, *348*, 399–408.

19. Kout, A.; Labed, S.; Chikhi, S.; Bourennane, E.B. AODVCS, a new bio-inspired routing protocol based on cuckoo search algorithm for mobile ad hoc networks. *Wirel. Netw.* **2018**, *24*, 2509–2519. [CrossRef]
20. Wang, F.; Chen, Z.; Zhang, J.; Zhou, C.; Yue, W. Greedy Forwarding and Limited Flooding based Routing Protocol for UAV Flying Ad-Hoc networks. In Proceedings of the 2019 IEEE 9th InternationalConference on Electronics Information and Emergency Communication, Beijing, China, 12–14 July 2018.
21. Choi, S.; Hussen, H.R.; Park, J.; Kim, J. Geolocation-Based Routing Protocol for Flying Ad Hoc Networks (FANETs). In Proceedings of the 2018 Tenth International Conference on Ubiquitous and Future Networks (ICUFN), Prague, Czech Republic, 3–6 July 2018.
22. Darabkh Khalid, A.; Alfawares, M.G.; Althunibat, S. MDRMA: Multi-data rate mobility-aware AODV-based protocol for flying ad-hoc networks. *Veh. Commun.* **2019**, *18*, 100163. [CrossRef]
23. Mahmud, I.; Cho, Y. Adaptive Hello Interval in FANET Routing Protocols for Green UAVs. *IEEE Access* **2019**, *7*, 63004–63015. [CrossRef]
24. Yang, H.; Liu, Z.Y. An optimization routing protocol for FANETs. *Eurasip J. Wirel. Commun. Netw.* **2019**, *2019*, 120. [CrossRef]
25. Zheng, Z.; Sangaiah, A.K.; Wang, T. Adaptive Communication Protocols in Flying Ad Hoc Network. *IEEE Commun. Mag.* **2018**, *56*, 136–142. [CrossRef]
26. Nayak, A.; Stojmenovic, I. *Wireless Sensor and Actuator Networks: Algorithms and Protocols for Scalable Coordination and Data Communication*; Wiley-Interscience: Hoboken, NJ, USA, 2010.

electronics

MDPI

Communication

An Improved Combination Method of MoM and UTD for Calculating the Radiation Characteristics of Antenna Arrays Mounted around Electrically Large Platform

Nan Wang *, Junzhi Liu, Qi Zhang and Yu Zhang

Shaanxi Key Laboratory of Large Scale Electromagnetic Computing, School of Electronic Engineering, Xidian University, Xi'an 710071, China
* Correspondence: wangnan@mail.xidian.edu.cn

Abstract: An improved method, namely, combined UTD with MoM, is studied to calculate the disturbed pattern of array antennas mounted around an electrically large platform. Each array unit is fed in turns in an MoM processor and then substituted into the UTD process to pursue the calculation of the disturbed pattern. When solving the current coefficient in MoM, the linear superposition principle is used to realize the separate feeding of each array unit. The proposed method can avoid unreliable UTD results when an equivalent phase center of the antenna is sheltered, and it can improve the RMS error from more than 15 dB to about 1 dB in the T-shaped example given. No large amount of computing resources are needed for the proposed method; the accuracy, efficiency and necessity are illustrated by examples given.

Keywords: array antenna; disturbed pattern; electrically large platforms; MoM; UTD

Citation: Wang, N.; Liu, J.; Zhang, Q.; Zhang, Y. An Improved Combination Method of MoM and UTD for Calculating the Radiation Characteristics of Antenna Arrays Mounted around Electrically Large Platform. *Electronics* **2022**, *11*, 3369. https://doi.org/10.3390/electronics11203369

Academic Editor: Giovanni Crupi

Received: 5 September 2022
Accepted: 12 October 2022
Published: 19 October 2022

1. Introduction

In computational electromagnetics, the Uniform Geometrical Theory of Diffraction (UTD) method is a highly efficient method dealing with electrically large problems. UTD method finds its application in problems of electromagnetic scattering and radiating appearing in complex environments [1–4]. In the popular UTD method, analytical patches can be used to construct large-scale urban environments to analyze electromagnetic problems and parametric surfaces such as Non-Uniform Rational B-Spline (NURBS), and can be used to model arbitrary curved environments to calculate problems such as electromagnetic scattering.

Array antennas are widely used in wireless communication and modern radar system due to their high power, high gain and fast beam scanning performances. When the array antenna is mounted on electrically large platforms, its pattern will be disturbed. When using the UTD method to analyze the disturbed radiation of the array antenna, hybrid methods are usually required [5–9] since UTD cannot deal with calculation of current distribution on complex antennas. These hybrid methods possess good precision in calculation; however, iterative procedures are always needed to modify matrix equations, which will lead to great increase in computational resource demand and rapid decrease in computational efficiency as the electrical size of antenna and platform increases.

In order to improve the computational efficiency, some modified hybrid methods are presented with practical precision maintained [10,11]. In the UTD method, electromagnetic fields are supposed to be carried by various kinds of rays traveling from the source point to the observe point. Thus, in these methods, the electromagnetic fields can be approximated as radiated by an equivalent point source which is usually located at the phase center of the actual antenna. Obviously, compared to mall antennas with a regular radiating element such as dipole or monopole, the practical array antenna is always complex and its phase center is hard to find; therefore, it is not proper to describe the antenna with one equivalent point. Moreover, because the position of the antenna on the platform is changeable, there is a chance that the phase center is occluded so that some of the observation points may have no rays to reach, which results in serious distorted or false calculated distribution of fields.

In this paper, High Order Moment Method (HOMoM) [12–16] is used to simulate the antenna, and the quasihybrid scheme in [11] is used to calculate the disturbed pattern of antenna arrays mounted on electrically large platforms. In order to avoid the influence of the phase center as well as keeping coupling information between array units, the linear superposition principle is used to realize the separate feeding of the array unit in solving the current coefficient by the moment method, and then the final electromagnetic field is obtained by the sum of importing each feeding unit in turns as source points into the UTD process. The impedance matrix in MoM is not modified while the array units are separately fed, but the coupling information between array elements is preserved.

The accuracy and efficiency of the proposed method are illustrated by example of the disturbed pattern of a shipborne array antenna, and the necessity is illustrated by a typical T-shaped structure, in which the RMS error is improved from more than 15 dB to about 1 dB. The presented method can be applied to deal with the radiation problem of array antennas mounted around electrically large platforms with good efficiency.

2. Improved Quasihybrid Scheme

The quasihybrid scheme used in this paper can be described briefly in Figure 1 as follows:

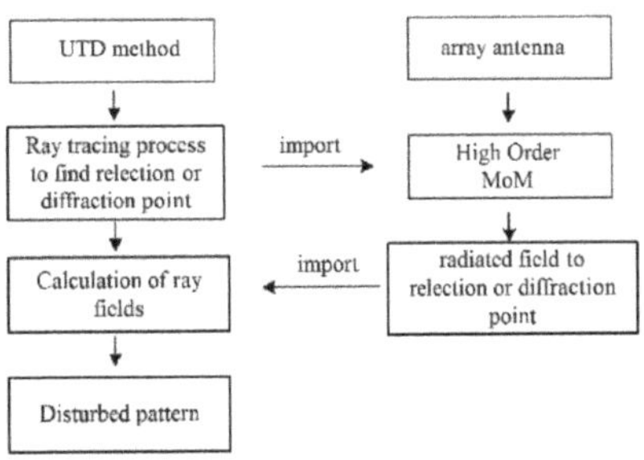

Figure 1. Brief procedure of quasihybrid scheme.

In this scheme, the antenna is modeled and calculated by MoM, and the resulting radiation field is then imported to the UTD method for further calculation.

In the MoM process, radiation of the antenna is treated by converting the electromagnetic equation into a matrix equation as follows:

$$[Z][I] = [V] \tag{1}$$

where $[V]$ is the excitation column vector and $[Z]$ is the impedance matrix of antenna units. When the current column vector $[I]$ is solved, the radiation of the antenna will be obtained.

Taking a dipole array antenna shown in Figure 2 as an example, Equation (1) can be written as:

$$\begin{bmatrix} Z_{11} & \cdots & Z_{1i} & \cdots & Z_{1n} \\ \vdots & \cdots & \vdots & \cdots & \vdots \\ Z_{i1} & \cdots & Z_{ii} & \cdots & Z_{in} \\ \vdots & \cdots & \vdots & \cdots & \vdots \\ Z_{n1} & \cdots & Z_{ni} & \cdots & Z_{nn} \end{bmatrix} \begin{bmatrix} I_1 \\ \vdots \\ I_i \\ \vdots \\ I_n \end{bmatrix} = \begin{bmatrix} V_1 \\ \vdots \\ V_i \\ \vdots \\ V_n \end{bmatrix} \tag{2}$$

where V_i is the excitation of unit i, Z_{ij} is the mutual impedance between unit i and unit j and I_i is the current coefficient of unit i to be calculated.

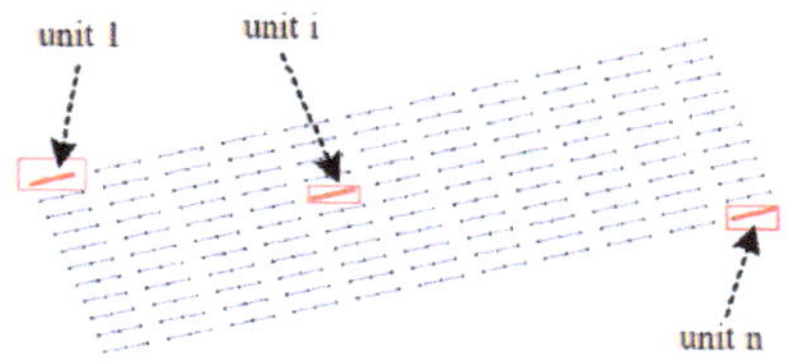

Figure 2. Illustration of a dipole array antenna.

In the UTD process, electromagnetic fields are supposed to be carried by various kinds of rays traveling from the source point to the observe point. These rays are approximated as transmitted from the equivalent point of the antenna array, which is usually located at the phase center or the geometrical center, as shown in Figure 3.

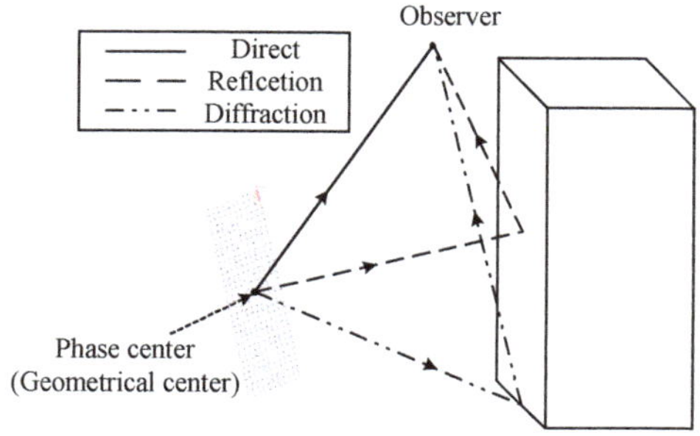

Figure 3. Fundamental ray types transmitted from the equivalent point of array antenna.

Obviously, compared to mall antennas with regular radiating element such as dipole or monopole, the practical array antenna is always complex, and its phase center or even geometrical center is hard to find; therefore, it is not proper to describe the antenna with one equivalent point.

Moreover, because the position of the antenna on the platform is changeable, there is a chance that the phase center is blocked so that some of the observation points may have no rays to reach, which results in a seriously distorted or falsely calculated distribution of fields. As can be seen in Figure 4, the path between the equivalent point and the observation point is blocked by the scatterer; thus, no field is transmitted. However, it can be seen clearly that the observation point is still in the area of the antenna aperture. A distorted result will be obtained under this condition.

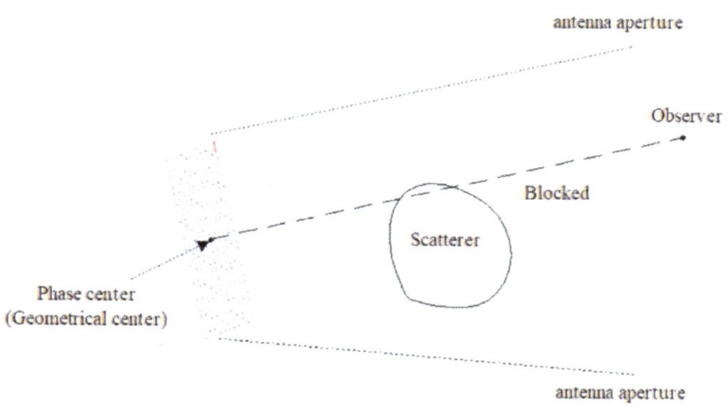

Figure 4. Drawback of the quasihybrid scheme dealing with array antenna.

To avoid the invalid calculation, the linear superposition principle is used to realize the separate feeding of each array unit in the MoM procedure.

Since the matrix equation in Equation (2) is a linear equation, if the excitation vector is split into the superposition of column vectors as in Equation (3), separate excitations for each unit can be achieved.

$$\begin{bmatrix} Z_{11} & \cdots & Z_{1i} & \cdots & Z_{1n} \\ \vdots & \cdots & \vdots & \cdots & \vdots \\ Z_{i1} & \cdots & Z_{ii} & \cdots & Z_{in} \\ \vdots & \cdots & \vdots & \cdots & \vdots \\ Z_{n1} & \cdots & Z_{ni} & \cdots & Z_{nn} \end{bmatrix} ([I_1] + \cdots + [I_i] + \cdots [I_n]) = \begin{bmatrix} V_1 \\ \vdots \\ 0 \\ \vdots \\ 0 \end{bmatrix} + \cdots + \begin{bmatrix} 0 \\ \vdots \\ V_i \\ \vdots \\ 0 \end{bmatrix} + \cdots + \begin{bmatrix} 0 \\ \vdots \\ 0 \\ \vdots \\ V_n \end{bmatrix} \quad (3)$$

where $[I_i] = \left[I_1^i \cdots I_i^i \cdots I_n^i\right]^T$ is the current column vector of the array when only unit i is excited by V_i, and they obey that:

$$\begin{bmatrix} Z_{11} & \cdots & Z_{1i} & \cdots & Z_{1n} \\ \vdots & \cdots & \vdots & \cdots & \vdots \\ Z_{i1} & \cdots & Z_{ii} & \cdots & Z_{in} \\ \vdots & \cdots & \vdots & \cdots & \vdots \\ Z_{n1} & \cdots & Z_{ni} & \cdots & Z_{nn} \end{bmatrix} \begin{bmatrix} I_1^i \\ \vdots \\ I_i^i \\ \vdots \\ I_n^i \end{bmatrix} = \begin{bmatrix} 0 \\ \vdots \\ V_i \\ \vdots \\ 0 \end{bmatrix} \quad (4)$$

I_j is the current coefficient of unit j when only the i-th antenna unit is excited by V_i, and this unit is brought into the UTD calculation process as the transmitting point. Together with the field of the whole array calculated based on Equation (4), Er_{MoM}^i, Ed_{MoM}^i, the ray field E_{UTD}^i at the observation point is then calculated by UTD process, as illustrated in Figure 5.

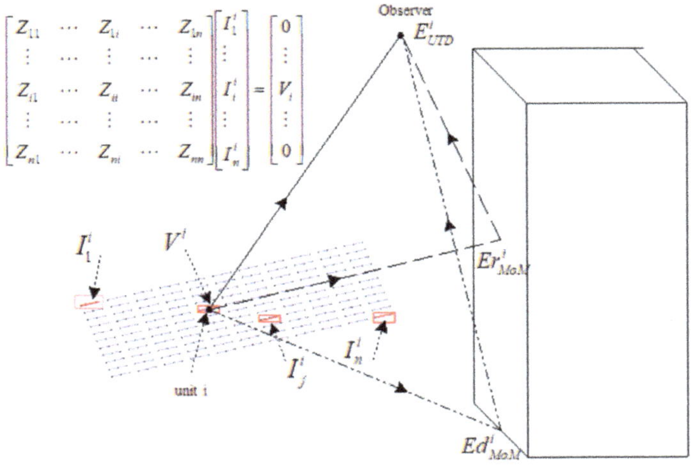

Figure 5. Illustration of calculation.

After all n units are excited in turn and the calculation is completed, the final radiation field can be achieved by the sum of ray fields calculated by the UTD method of each unit, $E = \sum_{i=1}^{n} E_{UTD}^i$.

There are two improvements in this calculation: 1. When only antenna element of unit i is excited, the coupling influence of other antenna elements is not ignored, and good calculation accuracy can be achieved while the calculation efficiency is preserved. 2. The radiation point of the antenna does not need to be equivalent to the phase center, and the calculation of the radiation field can still be effectively completed in the UTD procedure when the path between the phase center and the observation point is blocked.

3. Numerical Results

Practical examples are given in this section to illustrate the accuracy, efficiency and necessity of the proposed method.

Case 1. As shown in the Figure 6, an 8 × 8 dipole array antenna with the working frequency of 181.0 MHz is mounted on a ship structure. The ship is placed along the *z*-axis with a length of 150 m and a width of 16.5 m.

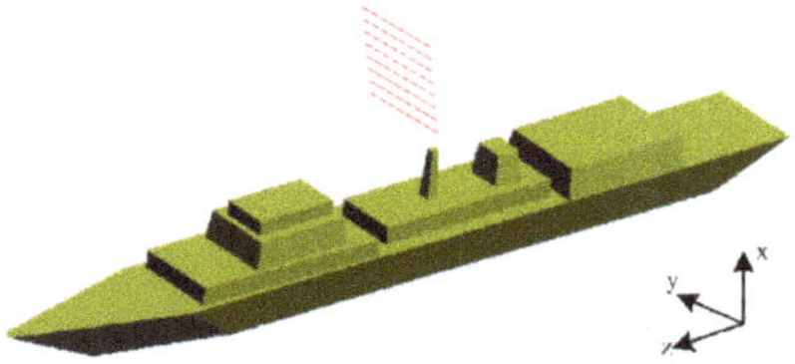

Figure 6. Illustration of shipborne array antenna.

The disturbed pattern on the yoz plane and xoz plane are calculated and compared to those calculated by HOMoM, as shown in Figure 7.

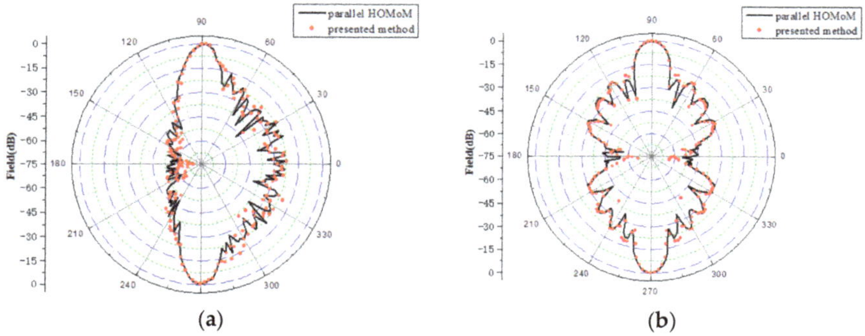

Figure 7. (**a**) Disturbed pattern on xoz plane; (**b**) disturbed pattern on yoz plane.

The requirements for computing resources of MoM are taken into account in this example; a parallel strategy is set and used on a high performance platform. It is necessary to point out that to obtain the disturbed pattern, the presented method does not need any high-performance resources beyond a common personal computer with 8 GB RAM. The cost of computing is listed and compared in Table 1, in which the efficiency of the proposed method can be seen.

Table 1. Comparison of computing cost.

	CPU Used	Memory Cost (GB)	Time Consumed (h)
HOMoM	576	918.1447	2.8687
Presented method	1	2.7246	1.2310

It can be seen from the results that the calculations are in good agreement. Additionally, it can be seen that the presented method can be applied to deal with the radiation problem of array antennas mounted around electrically large platforms with good efficiency.

Case 2. Shown in the Figure 8 is a T-shaped structure, which often appears on electrically large platforms such as aircraft tails or ship masts and is always located near antennas. A nine-dipole array antenna with the working frequency of 1.3 GHz is placed in front.

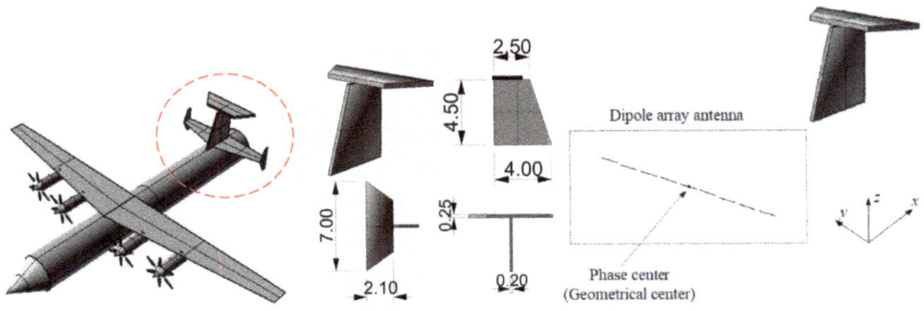

Figure 8. A T-shaped structure.

Firstly, the disturbed pattern on the xoz plane is calculated by the phase center-based quasihybrid method and compared to HOMoM, as shown in Figure 9. It can be seen clearly that serious distortion has appeared and been evaluated by RMS error; the numerical difference is as high as 16.9395 dB, which means that the result is no longer reliable. Such distortion is due to the fact that the phase center of the antenna is occluded at most observation points on the xoz plan. In the subsequent UTD calculation, the blocked phase center means the entire antenna is blocked, leading to a lack of field strength.

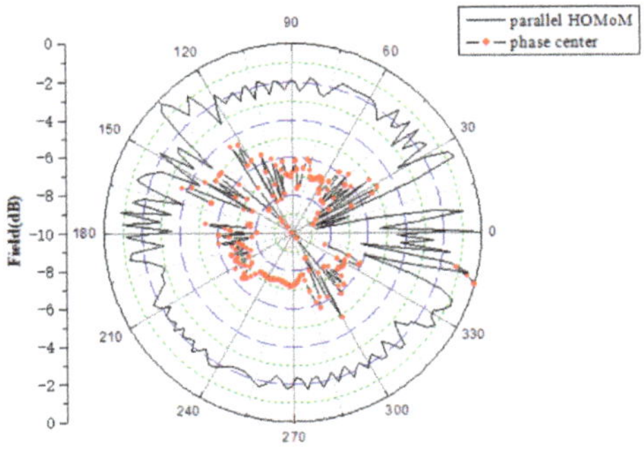

Figure 9. Disturbed pattern on xoz plane calculated by phase center.

Such calculation distortion can be prevented by using the presented method in this paper. Because each unit of the antenna is fed and imported into the UTD method as source point in turn, not all the antenna array is blocked; thus, the lack of field strength is eliminated.

Figure 10 shows the disturbed pattern on the xoz plane calculated by the method presented in this paper.

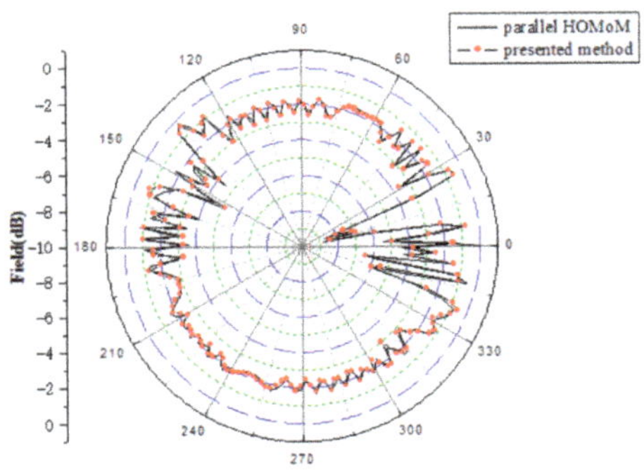

Figure 10. Disturbed pattern on xoz plane calculated by feeding each unit in turn.

It can be seen that the improved scheme can avoid the invalid calculation effectively and the RMS error is only 0.2758 dB in this example.

4. Conclusions

In this paper, an improved quasihybrid MoM–UTD method is used to calculate the disturbed pattern of an array antenna mounted around an electrically large platform. In the process of solving the current coefficient in MoM, the linear superposition principle is used to realize the separate feed of array units. Each feeding unit is then substituted in turns into the UTD process to pursue the calculation of the disturbed pattern. The presented method does not use one single equivalent source point to replace the antenna; therefore, the influences from the rest of the unfed unit can be considered simultaneously when a fed unit is used. Thus, good calculation accuracy can be maintained. Furthermore, the presented method can avoid the unreliable UTD results due to the shelter of one equivalent point, which can be seen in the example given where the RMS error from is improved from more than 15 dB to about 1 dB.

Author Contributions: Conceptualization, Y.Z. and N.W.; methodology, N.W.; validation, J.L. and Q.Z.; formal analysis, Q.Z.; investigation, J.L.; data curation, J.L.; writing—original draft preparation, Q.Z.; writing—review and editing, N.W. All authors have read and agreed to the published version of the manuscript.

Funding: This research received no external funding.

Conflicts of Interest: The authors declare no conflict of interest.

References

1. Tan, S.Y.; Tan, H.S. A Microcellular Communications Propagation Model Based on the Uniform Theory of Diffraction and Multiple Image Theory. *IEEE Trans. Antennas Propag.* **1996**, *44*, 1317–1326. [CrossRef]
2. Arenas, J.A.; Unai, U.; David, P.; Pablo, A.; Wout, J.; Luc, M. Statistical Analysis of Field Strength Location Variability for UHF Multimedia Broadband Services. *IEEE Antennas Wirel. Propag. Lett.* **2012**, *11*, 34–36. [CrossRef]
3. Daniele, V.; Lombardi, G.; Zich, R.S. Radiation and Scattering of an Arbitrarily Flanged Dielectric-Loaded Waveguide. *IEEE Trans. Antennas Propag.* **2019**, *67*, 7569–7584. [CrossRef]
4. Gennarelli, G.; Riccio, G. A Solution for the Plane Wave Diffraction by the Discontinuity Between Two Co-Planar Metamaterial Sheets. *IEEE Trans. Antennas Propag.* **2022**, *70*, 4705–4711. [CrossRef]
5. Civi, O.A.; Pathak, P.H.; Chou, H.T. An efficient hybrid UTD-MoM analysis of radiation/scattering from large truncated periodic arrays. *IEEE Trans. Antennas Propag.* **1998**, *2*, 748–751.
6. Bosiljevac, M.; Persson, P.; Sipus, Z. Efficient Analysis of Aperture Antennas on Generally Shaped Convex Multilayered Surfaces Using a Hybrid SD-UTD Method. *IEEE Trans. Antennas Propag.* **2009**, *57*, 1420–1431. [CrossRef]
7. Karagounis, G.; Zutter, D.D.; Ginste, D.V. A Hybrid MLFMM–UTD Method for the Solution of Very Large 2-D Electromagnetic Problems. *IEEE Trans. Antennas Propag.* **2016**, *64*, 224–234. [CrossRef]
8. Azpilicueta, L.; Falcone, F.; Janaswamy, R. Hybrid Computational Techniques: Electromagnetic Propagation Analysis in Complex Indoor Environments. *IEEE Antennas Propag. Mag.* **2019**, *61*, 20–30. [CrossRef]
9. Rodríguez, J.V.; Inglés, M.T.M.; Molina-García-Pardo, J.M.; Juan-Llácer, L.; Fujii, T.; Rodríguez-Rodríguez, I. UTD-PO Formulation for the Analysis of Multiple-Plateau Diffraction When Considering Illumination From a Low Source. *IEEE Trans. Antennas Propag.* **2021**, *69*, 4241–4245. [CrossRef]
10. Wang, N.; Dang, X.J.; Yuan, H.B.; Liang, C.H. Quasi-hybrid scheme of Non-uniform Rational B-Spline-Uniform Geometrical Theory of Diffraction Method for radiation of Antennas. *Microw. Opt. Technol. Lett.* **2011**, *53*, 2963–2967. [CrossRef]
11. Wang, N.; Yuan, H.B.; Xu, M.; Liang, C.H. Study on Interface Scheme for Non-Uniform Rational B-Spline Uniform Geometrical Theory of Diffraction Method. *Electromagnetics* **2016**, *36*, 392–399. [CrossRef]
12. Djordjevic, M.; Notaros, B.M. Double Higher Order Method of Moments for Surface Integral Equation Modeling of Metallic and Dielectric Antennas and Scatterers. *IEEE Trans. Antennas Propag.* **2004**, *52*, 2118–2129. [CrossRef]
13. Zhang, Y.; Lin, Z.C.; Zhao, X.W.; Sarkar, T.K. Performance of a Massively Parallel Higher-Order Method of Moments Code Using Thousands of CPUs and Its Applications. *IEEE Trans. Antennas Propag.* **2014**, *62*, 6317–6324. [CrossRef]
14. Lin, Z.C.; Gu, Z.J.; Zhao, X.W.; Zhang, Y.; Liu, H.W. An Efficient Matrix Equation Parallel Direct Solver for Higher-Order Method of Moments in Solution of Complex Electromagnetic Problems. *IEEE Access* **2018**, *6*, 29784–29792. [CrossRef]

15. Mao, T.; Lin, Z.; Zhao, X.; Wu, W. Electromagnetic Characteristics Analysis of Large Shipborne Array Antenna Based on Higher-Order MoM. In Proceedings of the 2020 International Conference on Microwave and Millimeter Wave Technology (ICMMT), Shanghai, China, 20–23 September 2020; pp. 1–3.
16. Zhang, C.; Cai, Q.-M.; Han, L.; Cao, X.; Zhu, Y.; Zhao, Y.-W.; Liu, T.; Fan, J. An Efficient Higher Order MoM-PO Method for EM Scattering From Electrically Large Objects. In Proceedings of the 2021 IEEE International Symposium on Antennas and Propagation and USNC-URSI Radio Science Meeting (APS/URSI), Singapore, 4–10 December 2021; pp. 1331–1332.

electronics

MDPI

Article

Effects of Introducing Confluence Rings on Ground Loss Resistance of VLF Umbrella-Type Antenna

Ying Quan, Hui Xie *, Cheng Yang, Hang Yu and Xinmiao Liu

College of Electronic Engineering, Naval University of Engineering, Wuhan 430033, China
* Correspondence: xiehui0723@163.com

Abstract: In this work, we analyze the influence of introducing confluence rings on the ground loss resistance in the ground grid system of the VLF umbrella-type transmitting antenna. The geometric deconstruction model of the confluence ring, the model of the VLF umbrella-type transmitting antenna, the models of the umbrella-type antenna ground grid system, and the formulae for the average conductivity are established. The working principle of the confluence ring was analyzed, and the ground loss resistance of the confluence ring structure with different numbers of layers and varying degrees of wire damage was simulated using Feko. The simulation results demonstrated the effect of increasing the number of confluence ring layers on the ground loss resistance of a fully functional grid to be relatively weak. However, when the ground grid wire is broken, the confluence ring structure can effectively reduce the ground loss resistance, thereby improving the radiation efficiency of the antenna.

Keywords: confluence rings; ground loss resistance; umbrella-type antenna; VLF

Citation: Quan, Y.; Xie, H.; Yang, C.; Yu, H.; Liu, X. Effects of Introducing Confluence Rings on Ground Loss Resistance of VLF Umbrella-Type Antenna. *Electronics* **2022**, *11*, 3419. https://doi.org/10.3390/electronics11203419

Academic Editor: Giulio Antonini

Received: 30 September 2022
Accepted: 18 October 2022
Published: 21 October 2022

1. Introduction

Very-low-frequency (VLF) communications use long wavelengths and a frequency range of 3–30 kHz for transmissions. The transmitting antenna is usually electrically small, with a low radiation resistance and a large ground loss resistance, resulting in inefficient radiation [1,2]. A large-scale ground network system is laid out in order to reduce the ground loss resistance and improve the radiation efficiency of the VLF shore-based transmitting antenna. The ground grid system of the VLF umbrella-type transmitting antenna is composed of ground grid wire, confluence ring, ground well, etc. (Figure 1). The confluence ring is a safety mechanism which acts as a bridge, electrically reconnecting the functional parts of a grid when a part of it breaks or corrodes, so as to avoid a sharp loss of resistance. Arthur Watt (1973) first elaborated on the calculation method for the ground loss resistance in the ground grid system [1]. Liu Chao et al. (2008) further expounded on the calculation of ground loss resistance for the umbrella-type antenna [2]. Furthermore, Ganlai Dai et al. (2018) analyzed and calculated the ground loss resistance of the large VLF umbrella-type antenna [3]. This was followed by the derivation of the equivalent admittance for the ground grid wire of the VLF umbrella-type transmitting antenna and the conductivity of the ideal and non-ideal ground grid wire modelby Li Bin (2019) [4].They established the models through Feko for simulations, with their predictions coming very close to the actual engineering results [4]. However, currently, there exists no literature on the calculation of conductivity for the ground grid system model (referred to as the ground grid system model). It is quite necessary to deduce this formula to improve the accuracy of the simulations to produce better predictions.

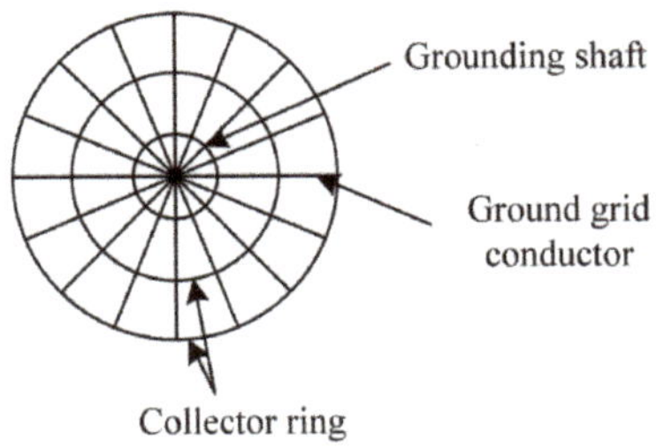

Figure 1. Structure of the ground network of the VLF umbrella-type transmitting antenna.

2. Theoretical Research on Confluence Ring and the Ground Grid System

2.1. Working Principle of the Confluence Ring

As shown in Figure 2a, the red region indicates the fully functional areas of the ground grid wire; the green area is the area where the ground grid wire cannot work normally due to corrosion or fracture; the blue area is the area where the ground grid wire cannot work normally due to the damage of the ground grid copper wire in the green area; and the inner network copper wire in the area around the green area connected with the confluence ring cannot work normally (in the antenna near field area, the current flow in the ground grid wire is from the distance to the antenna root). In the absence of the confluence ring, if the copper wire of the ground grid gets damaged at a certain point, it will, in turn, damage all the other ground grids connected to it. This greatly reduces the current conducting efficiency of this region, having a huge impact on the ground loss resistance. Through the analysis of Figure 2b,c, it can be seen that the damaged area associated with the ground grid gradually reduces after laying the confluence ring. If the two edges of the damaged ground grid area are attachedto the two layers of the confluence ring, the effect of the damage becomes nullified, minimizing its influence on the ground loss resistance.

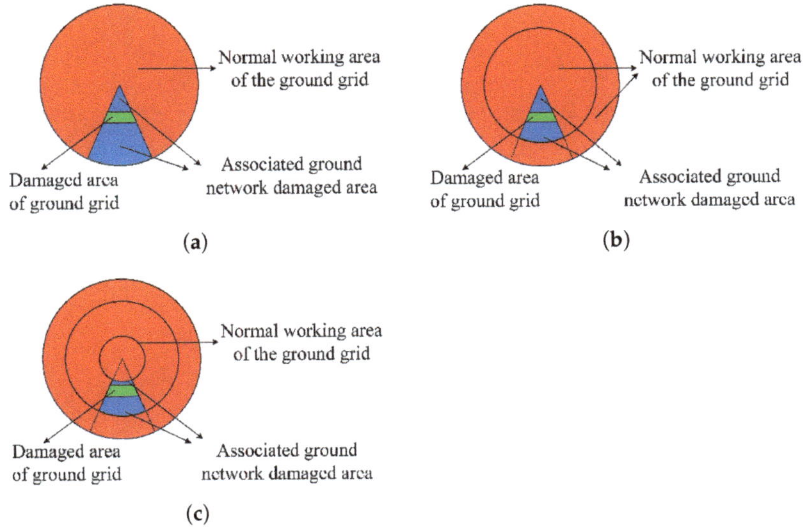

Figure 2. Schematic diagram of the different working conditions of the ground grid system. (**a**) In the absence of confluence rings; (**b**) after laying a few confluence rings; and (**c**) after laying multiple manifolds.

2.2. Geometric Deconstruction and Equivalent Conductivity Calculation of Flow Manifold Ring

The deconstructed geometric model of the confluence ring in the ground grid system is shown in Figure 3.

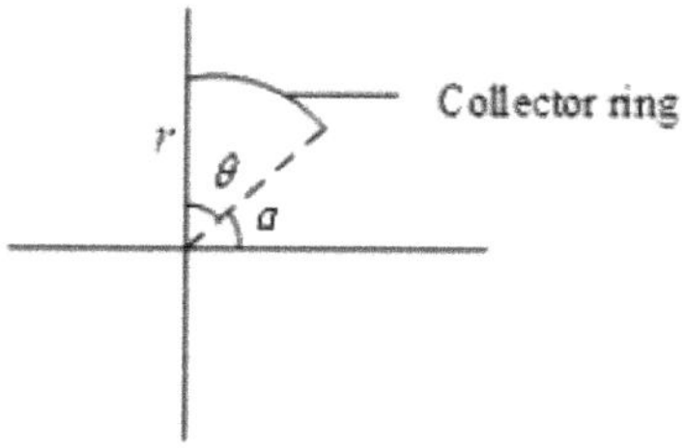

Figure 3. Geometric deconstruction of the confluence ring.

The vector magnetic potential of the confluence ring is given as

$$\vec{A} = \frac{\mu}{4\pi} Ir \frac{e^{jkl}}{l} d\theta \vec{e_\theta}, \tag{1}$$

where μ is the permeability of the confluence ring, I is the current on the confluence ring, k is the wave number, l is the arc length between two points on the confluence ring, r is the radius of the confluence ring, θ is the angle between two points on the confluence ring, and e_θ is the unit angle vector in the polar coordinates.

According to Biot–Savart law, the electric field generated by the confluence ring can be obtained as

$$\vec{E} = \frac{j\omega\mu}{4\pi} Ir \int \frac{e^{jkl}}{l} d\theta \vec{e_\theta}. \tag{2}$$

From Figure 2, we obtain

$$l = r\theta. \tag{3}$$

Combining Equations (1)–(3), we obtain [5,6]

$$\vec{E} = \frac{j\omega\mu\theta}{4\pi r(1 - jkr)} J(e^{jkr(\alpha+\theta)} \ln(\alpha + \theta) - e^{jkr\alpha} \ln \alpha)\vec{e_\theta}, \tag{4}$$

where J is the current density of the confluence torus.

Assuming that the confluence ring is made of an ideal conductor, due to the extremely long wavelength of VLF communication, the characteristic admittance of the non-ideal ground and the confluence ring is given as

$$Y = \sqrt{\left(\varepsilon_1 - \frac{j\sigma}{\omega}\right)/\mu_t} - j\frac{4\pi r(1 + k^2r^2)}{\omega\mu(\beta - \alpha)[\ln \beta(1 - k^2r^2\beta) - \ln \alpha(1 - k^2r^2\alpha)]}, \tag{5}$$

where ε_1 is the dielectric constant, σ denotes the electrical conductivity, and μ_t denotes the magnetic permeability of the soil.

$$L = \frac{4\pi r(1 + k^2r^2)}{\omega\mu(\beta - \alpha)[\ln \beta(1 - k^2r^2\beta) - \ln \alpha(1 - k^2r^2\alpha)]} \tag{6}$$

Let the equivalent characteristic admittance of the earth and the confluence ring be Y_α such that

$$Y_\alpha = \sqrt{\frac{\varepsilon_1 - \frac{j\sigma(r)}{\omega}}{\mu_t}}. \tag{7}$$

Now, Equations (6) and (7) can be used to obtain the formula for the equivalent conductivity of the Earth and the confluence ring structure as

$$\sigma(r) = \sigma + Re\left(\sqrt{\frac{\varepsilon}{\mu_t} + \frac{L^2}{4}} * (2L\omega\mu_t)\right). \tag{8}$$

Substituting the obtained results for those of study [3] for calculation, the formula for the average conductivity of the ground grid system model is obtained as

$$\sigma_e = \sigma + \frac{1}{b} * Re\left\{ \int_0^b \sqrt{\frac{\varepsilon}{\mu_t} + \frac{L^2}{4}} \right\} + \frac{1}{r_2 - r_1} \int_{r_1}^{r_2} \frac{\sqrt{\frac{2\sigma\omega}{\mu_t}}}{fdln\left(\frac{d}{2\pi\alpha}\right)} dr, \tag{9}$$

where r_1 and r_2 specify the ground network laying range, f is the operating frequency, ω represents the angular frequency, d is the arc length between adjacent ground network lines, a is the ground network wire diameter, and b is the confluence ring wire diameter.

2.3. *Umbrella-Type Antenna vs. Ground Grid System*

The VLF umbrella-type transmitting antenna model is established according to the Cutler antenna in the United States. The antenna structure parameters are as follows: the heights of the outer, inner, and center ring towers are 243 m, 266 m, and 298 m, respectively, and the radius of the inner and outer ring towers are 559 m and 960 m, respectively (Figure 4). Antenna sag is considered in the modeling process, and the wire is pulled. The wire conductivity is not considered [1,2], as the actual projected data from the past suggest that the wire loss resistance is essentially negligible compared to the total resistance. The support and radiation structure of the umbrella antenna are made of PEC. The top load of the umbrella is composed of six evenly distributed rhomboid antenna curtains, with a top capacitance line running on top of each curtain. The midpoint of each top capacitance line is connected to the by-pass lines of the two inner ring towers. The end points of the top capacity line are connected with the outer ring tower and the central tower, and the arc sag of the top capacitance line is 5%. Insulators are installed at the connections between all the top capacity lines and the inner ring tower, the outer ring tower and the central tower, and the insulators are replaced by 10 pF capacitors. The non-ideal geodetic conductivity was 0.01 S/m, the relative permittivity was 20, and the magnetic permeability was 3.6×10^{-3} H/m. Each antenna group is equipped with a radiant ground grid, buried at a depth of 0.5 m under the ground. Two hundred conductors with a length of 1200 m and a diameter of 0.004 m are uniformly laid, with operating frequencies at 15 kHz, 18 kHz, 20 kHz, and 24 kHz, respectively.

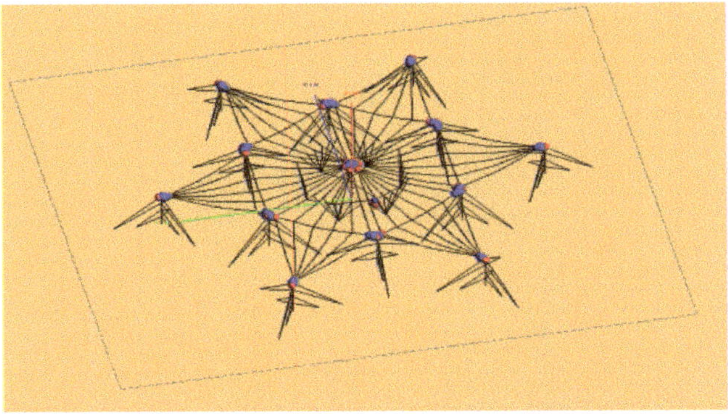

Figure 4. Model diagram of VLF umbrella-type transmitting antenna.

Eight different models were built in the FEKO software for each mentioned frequency. Firstly, under the condition that all the wires function normally, the ground grid system models with 2, 3, and 6 layers of confluence rings (referred to as the first type model) are uniformly established within a range of 1200 m, as shown in Figure 5.

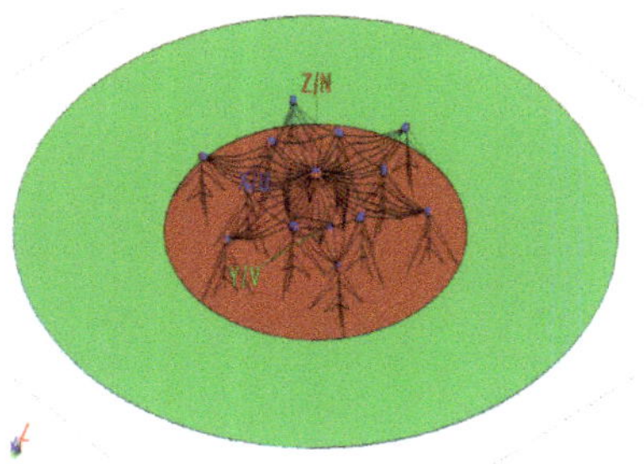

Figure 5. Schematic diagram of the first type model; the dark area is the ground network and the light area is the non-ideal ground.

With reference to Figure 3, establish a collector ring when laying the VLF umbrella antenna model (hereinafter referred to as the second category of the first model); a confluence ring when the VLF umbrella antenna model is not laid (hereinafter referred to as the second category of the second model); and the second category of the first model being laid refers to the convergence of the three layers of ring, with ground wire damage numbers of 10, 20, 30, 40 and 50. This is shown in Figure 6.

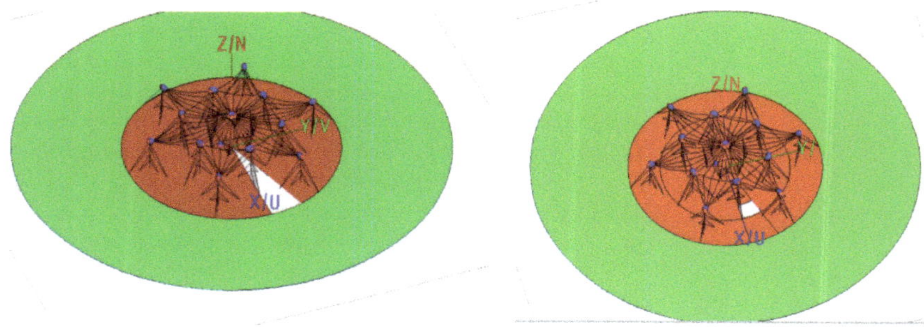

(**a**) (**b**)

Figure 6. Schematic diagram of the second type model; the darkly colored area is the functional part of the ground network, the lightly colored area is the non-ideal ground, and the blank area is the fracture in the ground network: (**a**) The first model. (**b**) The second model.

The two-type ground grid system model consists of a circular conductive medium plane with a thickness of 0.5 m. This plane is divided into three regions: (i) the functional and (ii) the damaged area of the ground grid wire, and (iii) the non-ideal ground. The dielectric constant and magnetic permeability of the conductive media in the former two regions are the same as that of the non-ideal ground. The conductivity of the damaged area of the ground grid wire and the non-ideal ground area is 0.01 S/m, and that of the functional area is then calculated using Equation (9).

3. Simulation Prediction and Analysis

3.1. Influence of the Number of Layers of Confluence Ring on Ground Loss Resistance

According to the antenna structure parameters in Section 2.3, a 3D VLF umbrella-shaped transmitting antenna model is simulated in Feko. The antenna is constructed using PEC, and the geodetic parameters corresponding to different models are set according to Section 2.3. In the calculation of antenna radiation resistance, the earth is considered to be an infinitely ideal earth. The non-ideal geodetic model is a concentric circle composed of different conductive media planes, and the radius of the outermost circle is the wavelength corresponding to the operating frequency of the antenna divided by 2π.

When the 200 ground grid two-layer, three-layer, and six-layer convergence ring models are laid, the average conductivity is calculated to be 0.0469 S/m, 0.0471 S/m, and 0.0477 S/m, respectively. These values were imported into the FEKO simulations, and the ground loss resistance of the first type model was calculated.

According to Figure 7, under the condition that all the ground grid wires are working normally, with the increase in the number of confluence ring layers, the earth loss resistance changes weakly, and the value change is less than 1%. It can be seen that the influence of the confluence ring on the ground loss resistance is not obvious when the ground grid wire is working normally.

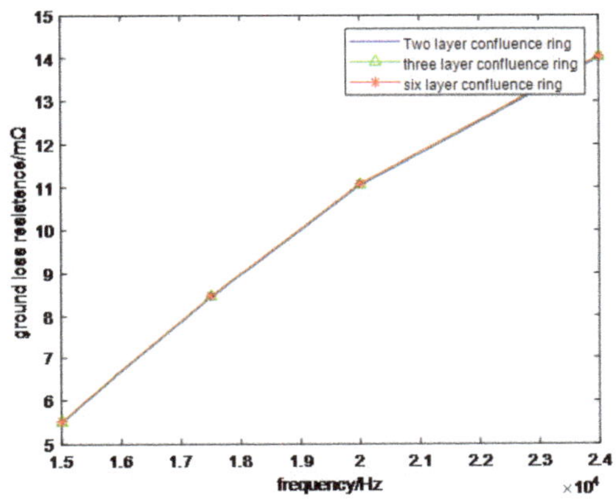

Figure 7. Ground loss resistors with different layers of manifold rings.

3.2. Influence of the Confluence Ring on the Ground Loss Resistance of Damaged Ground Grid Wires

For the second type of geodetic structure model, the calculation results of ground loss resistance are shown in Figure 8 and Tables 1 and 2. In case (A), all the ground grid wires are fully functional; in case (B), the ground grid wires are broken but the confluence ring is laid; and in case (C), the ground grid wires are broken and the confluence ring is not yet laid.

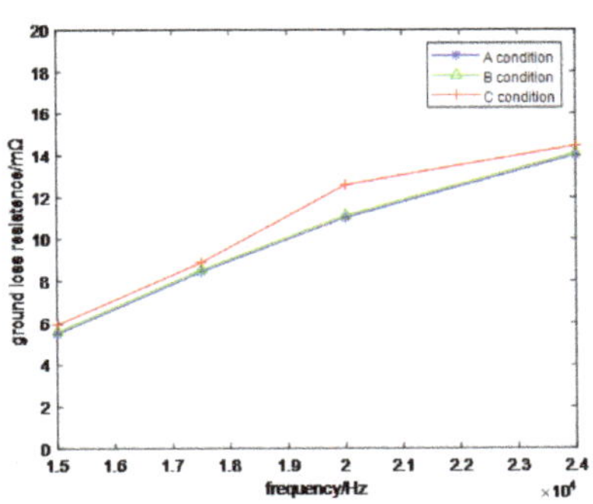

(**a**) Ground loss resistance under the condition of 10 broken grounding grids

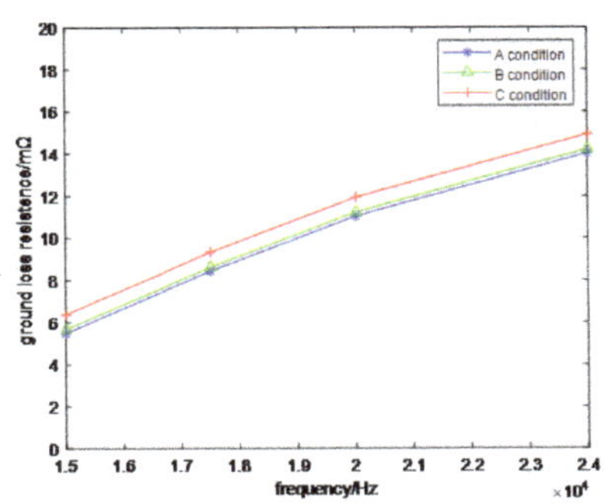

(**b**) Ground loss resistance under the condition of 20 broken grounding grids

Figure 8. *Cont.*

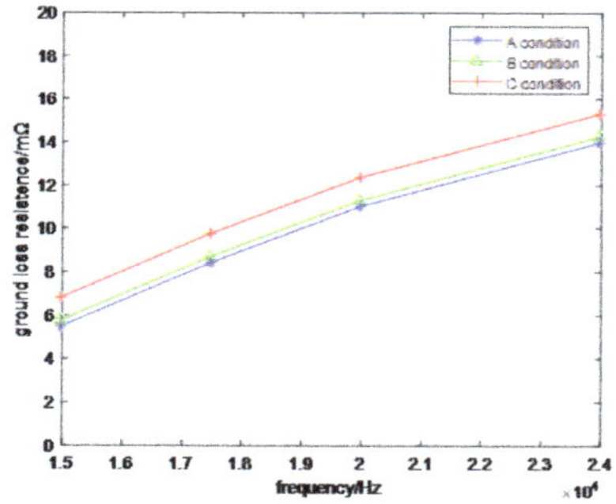

(c) Ground loss resistance under the condition of 30 broken grounding grids

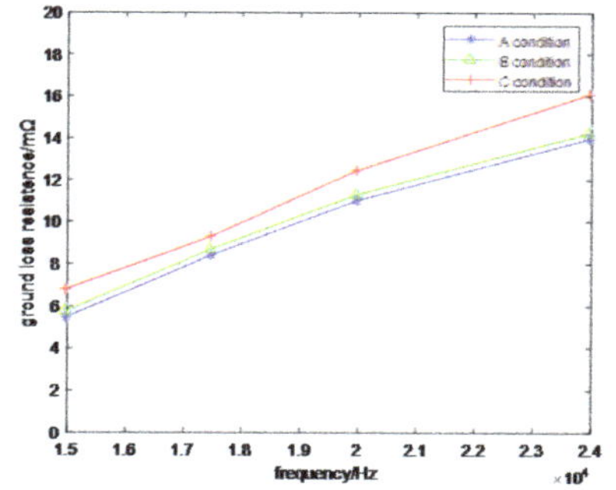

(d) Ground loss resistance under the condition of 40 broken grounding grids

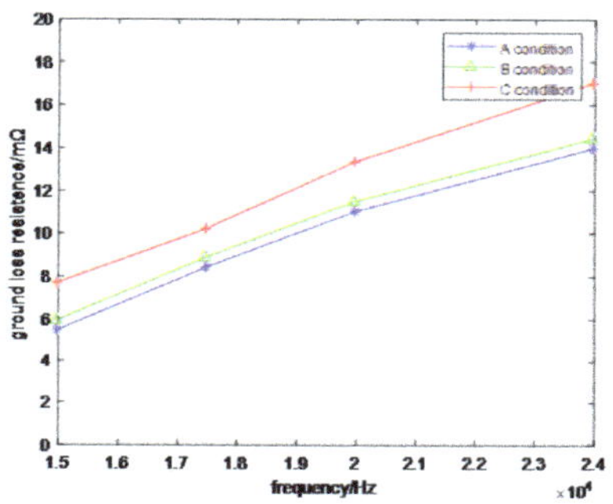

(e) Ground loss resistance under the condition of 50 broken grounding grids

Figure 8. Earth loss resistance under the three conditions A, B and C.

When the working frequency is 20 kHz, the percentage of improvement between the ground loss resistance of the second type of model and that of the corresponding fully functional ground grid (referred to as the normal resistance) is shown in Table 1.

Table 1. Percentage increase in earth loss resistance under B and C conditions.

	10	20	30	40	50
B	0.42%	0.9%	1.1%	1.6%	2.0%
C	4.2%	8.7%	12%	17.1%	20.4%

When the operating frequency is 20 kHz, the radiation efficiency of the second type of model antenna is shown in Table 2.

Table 2. The radiation efficiency of the antenna under two conditions B and C.

	10	20	30	40	50
B	70.67%	70.65%	70.62%	70.60%	70.58%
C	70.67%	70.54%	70.41%	70.28%	70.15%

According to Table 1, when compared to the normal resistance scenario, when 10, 20, 30, 40 and 50 ground grid wires are broken, the percentage increase in the resistance of case B was 0.42%, 0.9%, 1.1%, 1.6% and 2.0%, respectively, and the radiation efficiency was 70.67%, 70.65%, 70.62%, 70.60% and 70.58%, respectively. Whereas, for the same testing conditions, the percentage increase in the resistance of case C was 4.2%, 8.7%, 12%, 17.1%, and 20.4%, respectively, and the radiation efficiency was noted to be 70.67%, 70.54%, 70.41%, 70.28%, 70.15%, respectively. When the confluence ring is laid, the percentage increase in resistance is up to 2%, and that of radiation efficiency is down to 0.12% at most. Without the confluence ring, the percentage increase in resistance goes up to 20.4%, and that of radiation efficiency goes down to 0.52% at most. It is clear that when a multi-layer confluence ring is laid (which effectively reduces the damaged area associated with the ground grid), it reduces the ground loss resistance, and improves the radiation efficiency.

4. Summary

In this paper, the characteristic admittance of the ground grid confluence ring of the VLF umbrella-type transmitting antenna is derived, and the formula for the average conductivity of the equivalent model is obtained. The ground loss resistance is analyzed through Feko simulations for varying numbers of confluence rings and different degrees of damage to the wires. The following conclusions were drawn: (1) When the ground grid is functioning normally, the impact of the increasing number of the confluence ring layers on the ground loss resistance is minimal, and the percentage increase is far less than 1%. So, the effect of the confluence ring structure on the ground conductivity is not very obvious. (2) When the ground wire is damaged, the ground loss resistance increases with the number of damaged wires. In this case, introducing confluence ring layers into the ground grid system can substantially reduce the ground loss resistance and improve the antenna radiation efficiency.

Author Contributions: Conceptualization, Y.Q. and H.X.; methodology, Y.Q.; software, Y.Q.; validation, Y.Q., H.X. and C.Y.; formal analysis, Y.Q.; investigation, Y.Q.; resources, Y.Q.; data curation, Y.Q.; writing—original draft preparation, X.L.; writing—review and editing, X.L.; visualization, H.Y.; supervision, X.L.; project administration, Y.Q.; funding acquisition, Y.Q. All authors have read and agreed to the published version of the manuscript.

Funding: This research received no external funding.

Conflicts of Interest: The authors declare no conflict of interest.

References

1. Watt, A.D. *Very Low Frequency Radio Engineering*; Translation; National Defense Industry Press: Beijing, China, 1973.
2. Liu, C.; Jiang, H.; Huang, J. *Very Low Frequency Communication*; Haichao Publishing House: Beijing, China, 2008.
3. Dai, G.; Dong, Y.; Kang, Y. Analysis and Calculation of Ground Loss Resistance of large VLF Umbrella-type Antenna. *Commun. Technol.* **2018**, 33, 483–489.
4. Li, B. *Research on Radiation Characteristics of Very Low Frequency Current Antenna and Very Low frequency Umbrella-Type Antenna Array*; Wuhan Naval University of Engineering: Wuhan, China, 2019.
5. Feng, C.; Ma, X. *Engineering Electromagnetic Field*; Higher Education Press: Xi'an, China, 2011.
6. Jin, J. *Computational Electromagnetics*; Publishing House of Electronics Industry: Beijing, China, 2018.

electronics

Article

Improvement of the SPICE Model of Diode Based on Measurement and Nonlinear Fitting Random Optimization Algorithm

Dan Ren [1], Gang Xu [1,*], Jing-Qin Li [2], Ze-Yu Pan [2], Xuan Zhao [2] and Ping-An Du [2]

1 Institute of Electronic Engineering, China Academy of Engineering Physics, Mianyang 621999, China
2 School of Mechanical and Electrical Engineering, University of Electronic Science and Technology of China, Chengdu 611731, China
* Correspondence: xugangthu@163.com

Abstract: This paper proposes a modeling method for improvement of the SPICE Model of a diode based on measurement and a nonlinear fitting random optimization algorithm. First, the mechanism of electromagnetic interference generated by diodes at high frequencies was analyzed: dynamic characteristic parameters such as reverse recovery of diodes and junction capacitance. Second, the method of obtaining the I-V characteristic curve and junction capacitance characteristics of the diode through testing was introduced in detail, and a nonlinear fitting random optimization algorithm was proposed to calculate the static and dynamic characteristic parameters of the diode, which can be applied to improvement of the SPICE model of the diode. A simulation and test circuit were built, and the simulation and test results were compared in the time and frequency domains. The model in this study can not only truly reflect the peak voltage owing to the electromagnetic interference characteristics, but also ensure that the harmonic components and component amplitudes are within three times the fundamental frequency, verifying the accuracy of the electromagnetic compatibility model.

Keywords: diode; SPICE model; nonlinear fitting; simulation; verification

Citation: Ren, D.; Xu, G.; Li, J.-Q.; Pan, Z.-Y.; Zhao, X.; Du, P.-A. Improvement of the SPICE Model of Diode Based on Measurement and Nonlinear Fitting Random Optimization Algorithm. *Electronics* **2022**, *11*, 3461. https://doi.org/10.3390/electronics11213461

Academic Editors: Farhad Rachidi and Koichi Maezawa

Received: 19 August 2022
Accepted: 21 October 2022
Published: 26 October 2022

1. Introduction

In recent years, integration and miniaturization have been the main development directions of electronic products, which has made more and more electronic devices integrated in limited space, and brought serious electromagnetic compatibility problems into the system [1]. In order to solve the problem of electromagnetic interference in the system, electromagnetic compatibility (EMC) has gradually become one of the important concerns in the field of electronic system design. Since the 1960s, computational electromagnetism technology has developed rapidly, and many powerful computational electromagnetism algorithms and much software have been developed that make it possible to simulate EMC at device level, circuit level, board level and system level [2,3].

Diodes, as important nonlinear components, are widely used in various electronic systems and are important components considered in the EMC simulation. Leila et al. researched a new simplified model for predicting conducted EMI in DC/DC converters; the high-frequency disturbance estimation is based on the knowledge of circuit parasitic elements and semiconductor device parameters [4]. Kurt et al. focused on the modeling of bypass diodes in transient analysis and simulation photovoltaic installations [5].

Many studies have been conducted on diode modeling methods. Three ideal diode models that are widely used in engineering are: the ideal switching model, the constant voltage model and the segmented linear model [6]. However, these models are only suitable for simple circuit analysis, and the reason is that the junction capacitance and reverse recovery characteristics of diodes are not considered in these models, so they are not suitable for analyzing the electromagnetic interference characteristics of diodes.

Antognetti et al. proposed a model of a diode in the working state of DC and large and small signals, and studied the influence of temperature and area on the diode model parameters [7]. Deveney put forward a subcircuit model of an equivalent analog diode that focuses on the operation of the diode at high currents and over a certain temperature range, and builds a model that is compatible with most SPICE circuit simulators [8]. During the modeling of transient discharge suppressors, Bley et al. conducted an in-depth study of the nonlinear characteristics of the Zener diode, emphasizing that the capacitive effect needs to be considered when modeling the diode to reflect its high-frequency characteristics in practical applications [9]. Lepkowski et al. proposed a modeling method that fully considers the voltage–current relationship of transient voltage suppression diodes (TVS) for the circuit protection characteristics of transient voltage suppression diodes, focusing on the modeling principles and important physical characteristics, but did not carry out actual modeling work [10].

Piotrowski put forward a diode model that is compatible with the SPICE2 version. The simulation of the reverse leakage current region is optimized, and the parasitic series resistance of the reverse breakdown region is taken into account, and the feasibility of the model is verified by simulation calculations [11]. Zhang et al. simulated the charging and discharging process of resonant tunneling diodes in switching conversion through linear components such as resistive inductive capacitors, and at the same time use the second-order Longer-Kutta method to obtain the characteristic data of resonant tunneling diodes based on inductor fitting [12].

Tohlu et al. simulated the series resistance and junction capacitance characteristics of the diode through the approximate formula, and built a simulation platform to calculate the waveforms in different time domains, which verified the validity of the model [13]. Gan proposed a Zener diode equivalent circuit model based on resistors and voltage sources, which takes into account the dc characteristics to some extent, and improves the accuracy of circuit simulation, but does not fully consider the high-frequency characteristics such as junction capacitance and reverse recovery [14]. Hitchcock et al. established an equivalent circuit model of a Schottky diode based on the electrical parameters obtained by experimental tests, which is mainly composed of discrete components that characterize static and dynamics, and can simulate the operation of Schottky diodes at a certain temperature [15]. Yunyeong et al. proposed a LED SPICE model that focuses on the frequency and RC delay characteristics of the diode, and although the junction capacitance is affected by the frequency, the other characteristics of the diode are not fully analyzed [16].

Tao Liu et al. proposed a physical model method based on neural network acceleration to improve the traditional diode equivalent circuit model. The improved equivalent model can be used to analyze the operating state of the diode circuit under large injection and high frequency conditions, and the effectiveness of the model is verified by some examples [17]. A field-circuit coupling model for a merged PiN Schottky silicon carbide diode was proposed in [18], which considers the electrical characteristics of diode chip in the surge current condition and the influence of chip temperature coupling. The comparison between simulation and test results shows that the proposed field-circuit coupling model has the advantages of high simulation accuracy and efficiency.

Liu Qiang et al. modeled the physical characteristics of PIN diodes based on the finite element method, and applied the diode models brought by the Cadence software and the diode models built to the circuit simulation, respectively, and analyzed the differences in the electromagnetic interference characteristics [19]. Wu Chengkai studied the modeling method of the Schottky diode. Firstly, the theoretical calculation of the Schottky diode formation mechanism and model parameters is analyzed based on three-dimensional physical structure. Then, the influence of complex physical properties such as eddy current effect and proximity effect during operation of the Schottky diode on its performance is studied. Finally, the Schottky diode model is established and the optimization of the two cases is carried out, and the simulation results of the two optimization conditions are compared, and it is found that there is not much difference between the two. This model is only suitable for use in terahertz band multipliers and does not have extensive reference value for different types of diode EMC models [20].

As one of the important nonlinear components of the electronic system, the electromagnetic compatibility model parameters of the diode need to reflect its complex physical characteristics, and whether the electromagnetic compatibility model can be established accurately and efficiently will directly affect the effectiveness of the electromagnetic compatibility performance simulation of the electronic system. This study is based on the fast recovery diode 2CZ106C; we derive the key parameters of diode electromagnetic compatibility modeling from the functional characteristics of diodes, and propose a diode SPICE modeling method based on test data and the nonlinear fitting random optimization method. The SPICE model of 2CZ106C established in this paper not only accurately simulates the static characteristics of the diode, but also reflects the peak voltage generated by electromagnetic interference characteristics such as junction capacitance and reverse recovery, and guarantees the actual measurement and simulation of various harmonic frequencies within three times the fundamental frequency. The components and amplitudes of each component are consistent and can be used for electromagnetic compatibility simulations of electronic systems.

2. Analysis of the Mechanism of Diode Electromagnetic Interference

Due to the physical characteristics of the diode, such as junction capacitance and reverse recovery characteristics, it is easy to produce various complex harmonic interference to the working signal at high frequency, which is the cause of electromagnetic interference of the diode.

When the voltage at the diode terminal changes from forward to reverse, the current flowing through the diode will be reversed, but it is not immediately reduced to reverse saturation current; on the contrary, it remains basically unchanged for a period of time in the t_s, still in the reverse conduction state and not turned off. For PN junction diodes, the N-region storage hole does not disappear immediately, and the current gradually decreases only through the influence of the decimation of the reverse current and the recombination of the minority carriers (N-region hole).

After the period of t_s, the concentration of holes on the N-zone side of the PN junction is very small, and the storage charge in the N-region is almost completely gone. The gradient of the hole concentration at the boundary gradually decreases, and the reverse current begins to gradually decrease, and then after t_f, the current returns to I_s, and then returns to steady state. The above process is called the diode reverse recovery process, as shown in Figure 1.

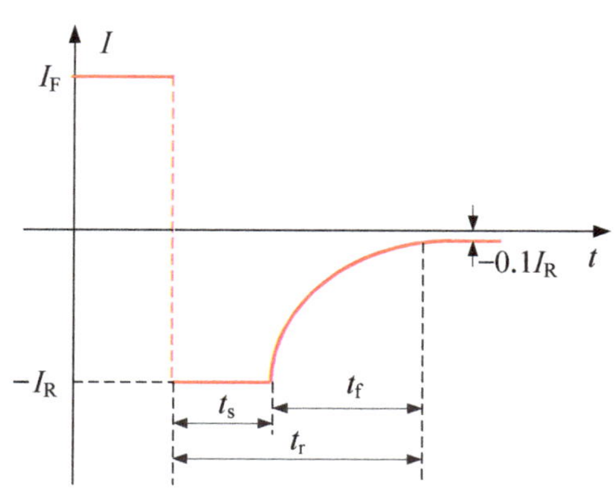

Figure 1. Diode reverse recovery process.

It is generally specified that after the t_s , t_f represents the time it takes when the current drops from $-I_R$ to $-0.1I_R$. Among them, I_R is the reverse current flowing through the diode when the applied voltage turns from forward to reverse during the reverse recovery measurement. t_s is called the storage time, t_f called drop time, $t_r = t_s + t_f$ called reverse recovery time or shutdown time [21].

From Figure 1, we can see that when the working signal of the electronic system changes from forward to reverse, due to this reverse recovery characteristic of the diode, the current flowing through the diode changes from forward to reverse, and the diode reverse current cannot be immediately reduced to zero to reach a complete cutoff state, which will lead to a very large di/dt. According to Maxwell's equations, this is the cause of the electromagnetic waves. Under the influence of parasitic parameters, these harmonic signals form inrush currents and peak voltages with short pulse periods and high interference intensities, resulting in interference. The diode reverse recovery time t_r can be expressed as:

$$t_r = \tau_p \ln\left[\frac{I_R + I_F}{I_R}\right] \tag{1}$$

In the formula, τ_p is the minority carrier lifetime.

When the applied voltage to the diode changes, the carrier distribution in the internal barrier region and the two neutral regions changes accordingly, thus creating a capacitive effect, called junction capacitance, which is generally of the order of pF, and decreases with the increase in the applied voltage. The higher the signal frequency, the more significant the junction capacitance effect, so the diode will generate a larger dv/dt for high-frequency signals, and even affect the rising and falling edges of the signal, resulting in conducted interference.

The junction capacitance determines the maximum operating frequency of a diode. When the signal frequency exceeds the maximum operating frequency, the diode will partially or completely lose unidirectional conductivity, resulting in signal distortion and harmonic interference at different frequencies. The diode junction capacitance can be expressed as:

$$C = \begin{cases} \dfrac{C_{J0}}{\left(1-\frac{V}{V_J}\right)^M} (V < FC\cdot V_J) \\ \dfrac{C_{J0}}{(1-FC)^{1+M}}\left[1 - FC(1+M) + M\frac{V}{V_J}\right], \ (V \geq FC\cdot V_J) \end{cases} \tag{2}$$

where V is the external bias voltage, FC is forward biased barrier capacitance, V_J, M, and C_{J0} are the built-in potential, gradient coefficient, and zero-bias barrier capacitance, respectively, and the junction capacitance can be calculated through these three parameters.

Ideally, the functional characteristics of the diode imply unidirectional conduction; that is, the forward low resistance is turned on and the reverse high resistance is turned off. The ideal positive I-V relationship varies exponentially, which can be expressed as:

$$I = I_S\left[exp\left(\frac{qV}{kT} - 1\right)\right] = I_S\left[exp\left(\frac{V}{V_T} - 1\right)\right] \tag{3}$$

The previous equation is the famous Schokley equation in semiconductor physics, V is the diode applied voltage, k is the Boltzmann constant, q is the electron charge, T is the absolute temperature, at room temperature $T = 300$ K, $V_T = \frac{kT}{q}$ is the temperature voltage equivalent, and at room temperature $T = 300$ K, there is $V_T = 26$ mV.

To consider factors such as the recombination of diode minority carriers in the barrier region, the ideal forward I-V relationship is modified in [22]:

$$I = I_S\left(e^{V/NV_T} - 1\right) \tag{4}$$

where I_S is the reverse saturation current, V_T is the voltage equivalent of temperature, which is 26 mV at room temperature. N is the emission coefficient that reflects the degree of correction of the actual forward I-V curve close to the ideal curve. The value of N, related to the material of the diode, generally needs to be obtained by measurement.

The ideal reverse I-V characteristic of a diode mainly considers the effect of the diffusion current. In addition, when the applied reverse voltage exceeds a certain value V_{BV}, the current exceeds the reverse breakdown current I_{BV} and rapidly increases exponentially, and the diode undergoes reverse breakdown and loses unidirectional conductivity.

In summary, diodes have dynamic characteristics such as reverse recovery and junction capacitance. At high frequencies, they generate various complex harmonics for the operating signal of the system. These harmonic signals cause conduction interference in the circuit. When the frequency is very high, it passes through the inductor. The coupling of parasitic parameters, such as the capacitance, can even cause radiated interference. At the same time, the nonlinearity of the diode I-V characteristic also affects the signal; therefore, the electromagnetic interference of the diode cannot be ignored.

3. Establishing a Diode EMC Model

3.1. Modeling Process

The diode SPICE model is an EMC model obtained from the perspective of the fundamental equations of semiconductor physics and the approximation of its parameters within a reasonable range. The SPICE model of the diode is mainly composed of two parts: the model equation and the model parameters. The SPICE simulator has established corresponding physical model equations for 16 basic component types and nested them in the software simulation engine. Model parameter files need to be called from the simulation software model library or imported externally, and combined with the simulator for simulation calculations.

The diode EMC model based on ".MODEL" mode mainly includes the 14 model parameters shown in Table 1, of which some model parameters have specific default values or typical values.

Table 1. The parameters of diode EMC model in ".MODEL" mode.

Number	Model Parameters	Default Value	Typical Value	Unit
1	Reverse saturation current I_S	1×10^{-14}	1×10^{-14}	A
2	Emission coefficient N	1	-	-
3	Series resistance R_S	0	10	Ω
4	reverse breakdown voltage V_{BV}	∞	50	V
5	reverse breakdown current I_{BV}	1×10^{-10}	-	A
6	Zero Bias Junction Capacitance C_{J0}	0	2 pF	F
7	gradient coefficient M	0.5	0.5	-
8	Built-in potential V_J	1	0.6	V
9	transit time TT	0	0.1 ns	s
10	flicker noise index AF	1	-	-
11	flicker noise figure KF	0	-	-
12	Band gap EG	1.11	1.11	eV
13	Forward Biased Barrier Capacitance FC	0.5	0.5	-
14	Reverse saturation current temperature index XTI	3	3	-

In general, the 14 model parameters in Table 1 can be divided into three categories: (a) There are static model parameters that reflect the DC characteristics and functional characteristics of the diode: I_S, N, R_S, V_{BV}, I_{BV}. Among them, I_S, N and R_S characterize the forward bias characteristics of the diodes, V_{BV} and I_{BV} characterize the reverse bias characteristics; (b) The dynamic model parameters that reflect the electromagnetic interference characteristics of diodes are C_{J0}, M, V_J, TT, where C_{J0}, M, V_J characterize the relationship between junction capacitance and reverse bias voltage. TT describes the reverse recovery characteristics, which are affected by the peripheral test circuitry; (c) The model parameters determined by the process or semiconductor material are XTI, KF, AF, EG and the positive partial depletion capacitance correction coefficients FC that are not affected by other parameters are also attributed to this category, and it is generally believed that according to experience or process materials, they can be directly determined, as known quantities do not need to be optimized.

In the establishment of a comprehensive and accurate diode electromagnetic compatibility model, it is necessary to obtain all the model parameters in Table 1, but usually the diode parameter manual can only provide a part of the parameters; for the model parameters that the parameter manual cannot provide, most of the cases need to obtain the curve of the parameters through testing, and then are further based on the fitting algorithm, the model parameter extraction of the test curve.

The model parameters of the diode junction capacitance characteristics, reverse recovery characteristics and I-V characteristics are derived from the electrical characteristic parameters and curves, which can be obtained by fitting the measured data. An experimental platform is built to test the time-domain waveform of the diode reverse recovery process, so the forward current I_F, the reverse current I_R, reverse recovery time t_r, and then the reverse recovery characteristics TT can be calculated by Equation (1). In addition, the reverse saturation current temperature index XTI, flicker noise figure KF, flicker noise index AF, forbidden bandwidth EG and forward bias barrier capacitance coefficient FC are usually classified as model parameters determined by material technology, generally based on semiconductor theory or material technology, which can be directly determined.

After obtaining all the parameters, the diode EMC model can be established according to the SPICE grammar rules. The diode EMC modeling method we propose is shown in Figure 2. The EMC model of the diode obtained by this paper can not only reflect the functional characteristics and electromagnetic interference characteristics of a diode, but also be suitable for computational electromagnetic simulation, which can provide effective technical means and construction ideas for the establishment of various diode EMC models.

3.2. Nonlinear Fitting Stochastic Optimization Algorithm

As mentioned above, in general, the diode's parameter manual does not provide all the parameters needed for modeling, and these missing parameters can only be obtained by the measurement and then by the data fitting method. In this paper, a nonlinear fitting stochastic optimization algorithm is established based on the diode physical characteristic equation, combined with the Nlinfit or Lsqcurvefit function. The algorithm combines Equations (2) and (4) to fit and calculate the model parameters of the I-V characteristics and junction capacitance characteristics and obtain the reverse saturation current I_S, emission coefficient N, zero-bias junction capacitance C_{J0}, gradient coefficient M and built-in potential V_J. The principle of the optimization algorithm is shown in Figure 3.

The basic idea of the algorithm is random fitting optimization based on the idea of least squares, so that the fitting objective function (diode physical equation) continues to approach the test data until the residual sum of squares reaches the set accuracy or the number of fittings reaches the loop setting. At the maximum value, the algorithm terminates the calculation and finally obtains the optimal solution of the model parameters. We consider setting the rough residual sum of squares to improve the optimization direction, so that the fitting results gradually approach the target accuracy.

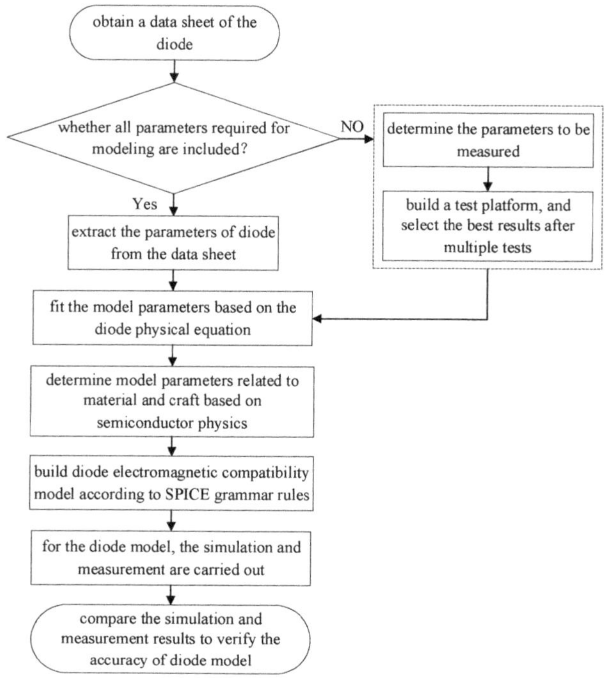

Figure 2. Flow chart of diode EMC modeling method.

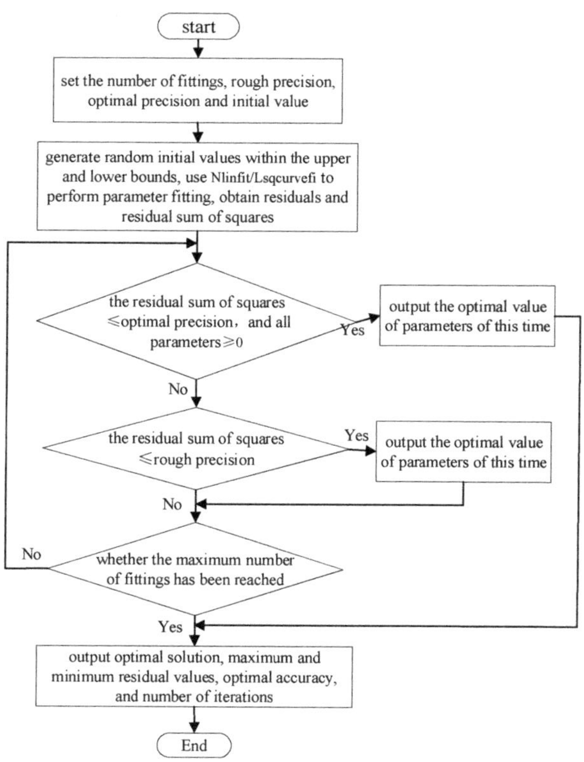

Figure 3. Flow chart of nonlinear fitting stochastic optimization algorithm.

Because the initial value of the model parameters will affect the fitting speed and solution results, it is necessary to provide a reasonable initial value to obtain the optimal solution of the model parameters. The specific initial value solution method used in this study is as follows:

$$x_0 = x_1 + b \times (x_h - x_1) \tag{5}$$

where, x_0, x_h, x_1 are the initial value, the upper and lower limits of the limited initial value range, and b is a random number in the range of (0, 1) generated by the rand function.

According to the test data, the approximate derivative values of two or three groups of specific points can be obtained approximately, and further combined with the first-order derivatives of Equations (2) and (3) to obtain V_J, M, C_{J0}, N and I_S, a total of five model parameters, considering the result set as the upper limit of the initial parameter value in the fitting optimization algorithm. In addition, the original default value of the diode SPICE model parameters is selected as the lower limit of the initial value of the fitting algorithm. The influence of the test data on the fitting effect is preliminarily considered, and the reference value of the default value of the diode SPICE model parameters is fully reflected.

3.3. The Measurement of Diode

The electrical performance test of the diode is divided into static electrical parameter test and dynamic electrical parameter test; the static electrical parameter is mainly described by the I-V curve, and the dynamic electrical parameter mainly considers the junction capacitance and reverse recovery characteristics.

The I-V static parameter test system for diodes is shown in Figure 4, which consists of the Agilent 4155B semiconductor parameter analyzer, a semi-automatic probe test bench, and a computer with Agilent EasyExpert software installed. The Agilent 4155B semiconductor analyzer has a current-voltage resolution of 10 fA/µV and a corresponding measurement range of 100 mA/100 V. Agilent EasyExpert test software provides the Agilent 4155B with a full range of diode static parameter testing that controls test instruments and probe stations for fully automated I-V scanning tests in DC or pulse mode. The test layout of the 2CZ106C fast recovery diode is shown in Figure 5.

2CZ106C is an N-type silicon rectifier diode, and the forward conduction voltage of the silicon diode is about 0.5~0.7 V. Therefore, when testing the forward I-V curve, the set voltage test range is 0~1.5 V, and the current test range is 0~100 mA. In addition, due to the overvoltage protection mechanism of the Agilent EasyExpert test software, the maximum voltage value can be set to −50 V and the maximum measured current value is −10 mA. Therefore, when testing the reverse I-V curve, the reverse test voltage range is −50~0 V, and the test current range is set to −10~0 mA.

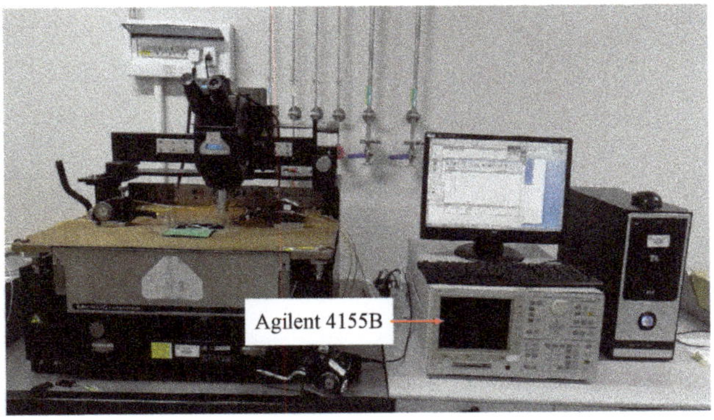

Figure 4. The I-V static parameter test system for diode.

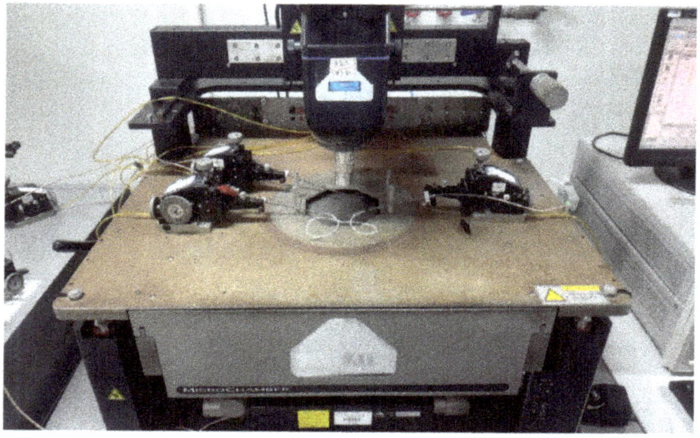

Figure 5. The I-V static parameter test layout of the 2CZ106C.

After completing the test system connection, test calibration via Agilent EasyExpert software is also required to eliminate test instrument errors and ensure the accuracy of experimental data. The forward I-V test results are shown in Figure 6, and it can be seen from Figure 6 that at about 1.15 V forward bias, the forward current of the diode 2CZ106C reaches the upper limit current of 100 mA allowed by the instrument, and its opening voltage is about 0.5 V, and the on-voltage is about 0.6~0.7 V. The test results of reverse I-V are shown in Figure 7, where it can be seen that the reverse leakage current of diode 2CZ106C is nA level within the reverse bias voltage of −50 to 0 V (that is, before reverse breakdown).

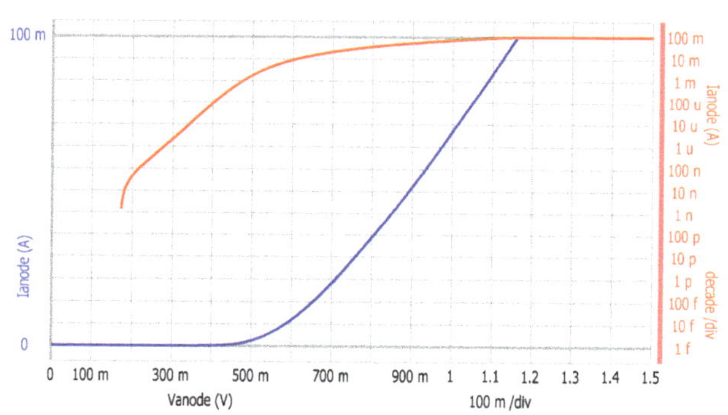

Figure 6. The forward I-V characteristic curve test results of the 2CZ106C.

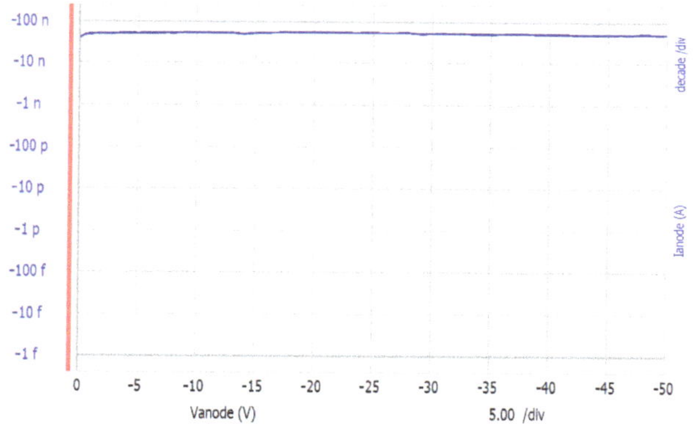

Figure 7. The reverse I-V characteristic curve test results of the 2CZ106C.

The forward I-V curve of the diode is parameterized by using the Lsqcurvefit nonlinear random optimization algorithm to obtain the optimal solution of the reverse saturation current and emission coefficient, the absolute values of the maximum and minimum residual differences, the sum of squares of the residuals with the best accuracy and the maximum number of iterations as shown in Table 2.

Table 2. Parameter fitting result of the forward I-V characteristic model of the diode.

Number	Optimization Parameters	Value	Unit
1	Maximum number of cycles	10,000	time
2	Reverse saturation current I_S	4.377251×10^{-6}	A
3	Emission coefficient N	3.087715	-
4	The sum of squares of the residuals of the optimal accuracy	6.302667×10^{-5}	A2
5	The absolute value of the maximum residual	1.468315×10^{-3}	A
6	The maximum residuals correspond to the voltage value	0.6585	V
7	The absolute value of the minimum residual	2.189795×10^{-6}	A
8	The minimum residuals correspond to the voltage value	0.534	V

For nonlinear random optimization fitting, the difference between the fitted value and the measured value is called the residuals, and the sum of all the residuals squared can be used to measure the goodness of the fit as a whole, and if the value is a very small order of magnitude, close to zero, the fit is good. From the results of Table 2, when the maximum number of cycles = 10000 is reached, the reverse saturation current $I_S = 4.377251 \times 10^{-6}$ and the emission coefficient $N = 3.087715$. The parameter fitting of the Lsqcurvefit nonlinear stochastic optimization algorithm fails to achieve the target optimal residual squared sum and accuracy rsk = 1×10^{-10} originally set by the program, and the actual obtained residual squared sum of squares is 6.302667×10^{-5}. In addition, the absolute value of the maximum current residuals obtained at 0.6585 V of the measured data used is 1.468315×10^{-3} A, and the absolute value of the minimum residual current obtained at a voltage of 0.534 V is 2.189795×10^{-6} A.

Figure 8 shows the comparison between the fitted curve and the measured curve. It can be seen from Figure 8 that the fitted curve is basically consistent with the measured curve; the two curves are well matched. Therefore, the diode parameters obtained by fitting have high accuracy.

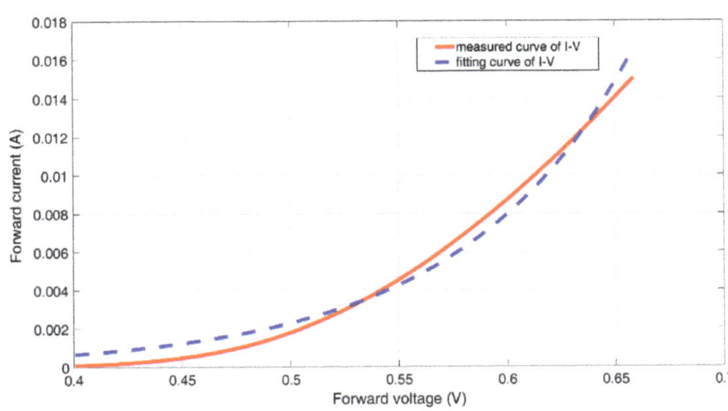

Figure 8. The comparison between the fitted and measured curve of I-V.

As mentioned earlier, the dynamic parameters of the diode focus on the junction capacitance and reverse recovery time parameters.

The Agilent 4294A Precision Impedance Analyzer can test components and circuits for electrical characteristics such as impedance curves and junction capacitance curves, and the test accuracy can reach pF level. The Agilent 4294A impedance analyzer is built with the Signaton S-1060 series probe test bench and a computer equipped with the test sharing software QuickTest to form a semiconductor device test system as shown in Figure 9, and the junction capacitance characteristic curve of the diode 2CZ106C is tested experimentally.

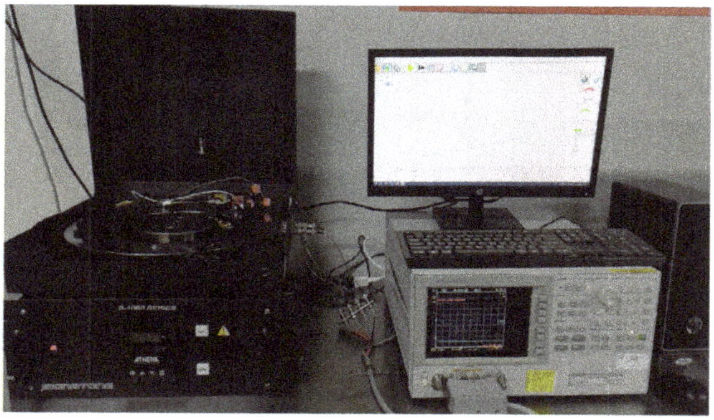

Figure 9. The junction capacitance parameter test platform for diodes.

Since the order of magnitude of diode junction capacitors is generally pF level, which is a smaller capacitance, when testing the diode 2CZ106C junction capacitance, we choose a $C_p - R_p$ combination of measurement parameters in parallel. The test step is to open the Agilent 4294A impedance analyzer in advance to warm up for at least 20 min, then perform phase compensation, open circuit compensation, short circuit compensation and load compensation, open the test software QuickTest, set the sample information and select the bias parameter for testing. According to the actual environmental conditions, the test temperature is set to 19.6 °C, the sweep bias frequency is 1 MHz, the DC bias range is −40~0 V and the test data point is 26. Only a bias sweep C_p of the diode 2CZ106C junction capacitance curve data can be obtained, so the test parameters are selected C_p and not displayed R_p. The detail of the diode's junction capacitance parameter test is shown in Figure 10, and the test result is shown in Figure 11.

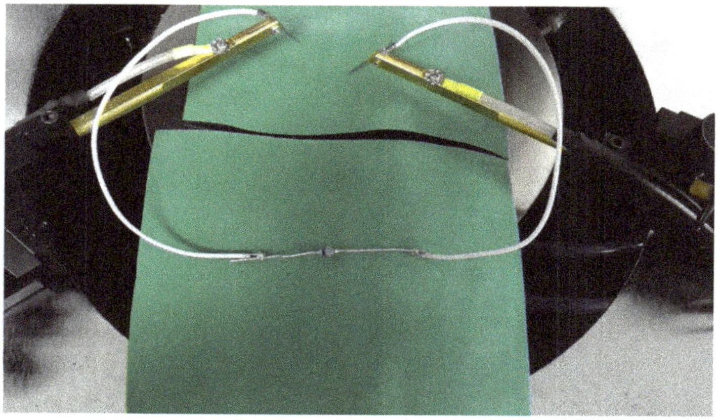

Figure 10. The detail of the diode's junction capacitance parameter test.

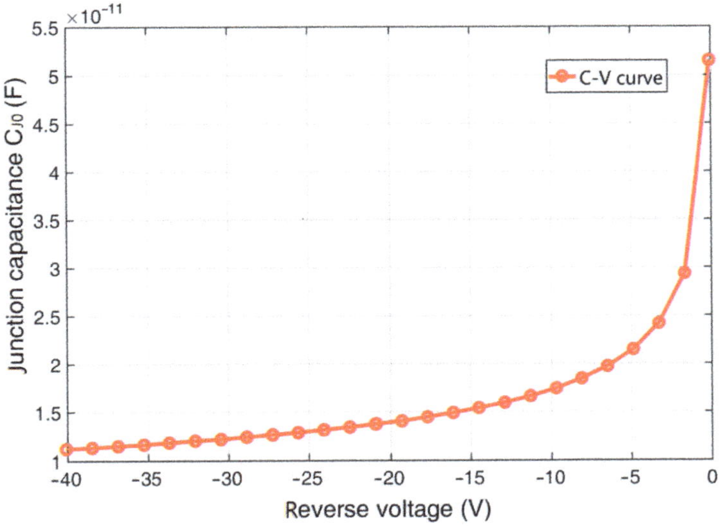

Figure 11. The junction capacitance test result.

From Figure 11, we find that the diode 2CZ106C junction capacitance decreases with the increase in the applied reverse voltage, and the order of magnitude is generally pF. From this qualitative analysis, the test results basically reflect the characteristics of the diode junction capacitance.

Considering the junction capacitance of the diode in the case of $V < FC{\cdot}V_J$ as the objective function, and Nlinfit nonlinear random optimization algorithm is applied to the fitting of diode junction capacitance characteristic model parameters, and the optimal solution of built-in potential V_J, gradient coefficient M, zero bias junction capacitance C_{J0}.

First, the derivative values of three groups of specific points are estimated from the experimental data, and then substituted into the first-order derivative expression of the fitting objective function. We observe the values of V_J, M and C_{J0} are 1.648×10^{-11}, 16.545 and 1.636, respectively. The analytical solutions of V_J, M and C_{J0} are considered as the upper limit of the initial value in the nonlinear fitting stochastic optimization algorithm, and the typical values are set as the lower limit of the initial value of the fitting.

The absolute value of the maximum and minimum residuals, the sum of the residuals of the optimal accuracy, the maximum number of iterations and other information are shown in Table 3.

Table 3. Parameter fitting result of the C-V characteristic model of the diode.

Number	Optimization Parameters	Value	Unit
1	Maximum number of cycles	10,000	time
2	zero bias junction capacitance C_{J0}	5.150461×10^{-11}	pF
3	built-in potential V_J	0.319813	V
4	gradient coefficient M	0.315106	-
5	The sum of squares of the residuals of the optimal accuracy	2.049740×10^{-26}	pF2
6	The absolute value of the maximum residual	7.765136×10^{-14}	pF
7	The maximum residuals correspond to the voltage value	1.6	V
8	The absolute value of the minimum residual	5.582152×10^{-16}	pF
9	The minimum residuals correspond to the voltage value	20.8	V

As can be seen from Table 3, when the maximum number of cycles t = 10,000 is reached, three model parameters of the diode are obtained: built-in potential V_J = 0.319813, gradient coefficient M = 0.315106 and bias junction capacitance C_{J0} = 5.150461 × 10^{-11}. The parameter fitting calculation fails to achieve the original target optimal residual squared sum accuracy rsk = 1 × 10^{-30} pF^2, and the actual residual squared sum and the optimal accuracy are 2.049740 × 10^{-26} pF^2. In addition, the absolute value of the maximum residual value of the capacitor is obtained at a voltage of 1.6 V, S_{max} = 7.765136 × 10^{-14} pF, and the minimum residual absolute value S_{min} = 5.582152 × 10^{-16} pF is obtained at a voltage of 0.534 V. Both the absolute value of the maximum residuals and the optimal accuracy of the actual residuals are very close to zero, indicating that the curve fitting has a high degree of accuracy.

Figure 12 shows the result of comparing the fitted curve of the junction capacitance characteristics with the measured curve. From the figure, it can be seen that the fitted curve is exactly the same as the measured curve.

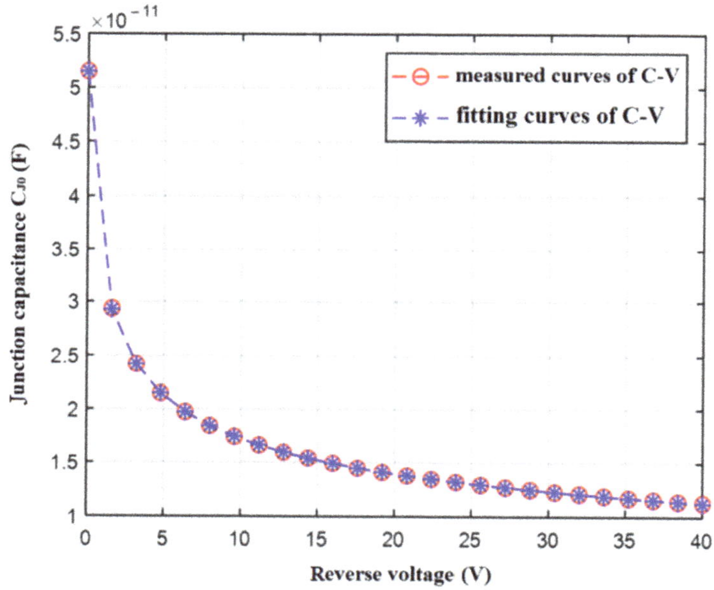

Figure 12. Comparison of fitting and measured curves of junction capacitance characteristics.

In diode EMC models, reverse recovery characteristics are generally characterized by time-of-flight TT. The reverse recovery time t_r of diode 2CZ106C can be obtained by building an external circuit test, and the model parameter TT can be further obtained. For the experimental test of diode reverse recovery time t_r, the square wave with a given duty cycle of 50% can be used as the excitation signal by the F40-type digital synthesis function signal generator. We use the Keysight MSOX2024A mixed-signal oscilloscope to test the time-domain waveform at both ends of the load to obtain the diode reverse recovery process. At the same time, through the two longitudinal wiper of the oscilloscope, the reverse recovery time of the diode is accurately read, and the transition time TT is solved using the physical formula of the previous section and the test platform is shown in Figure 13.

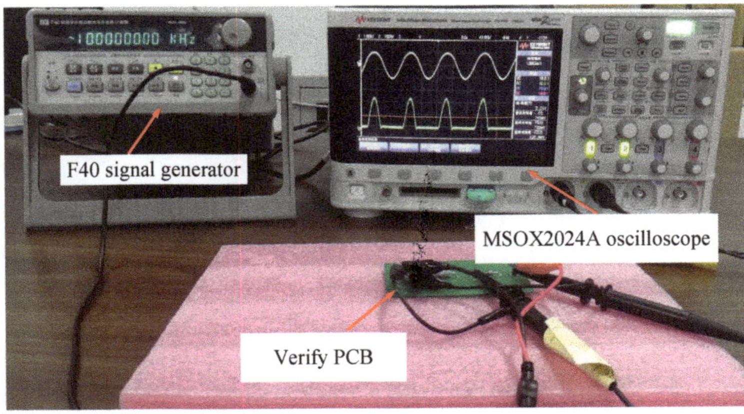

Figure 13. The reverse recovery characteristics test system for diode.

A square wave signal with a frequency of 1 kHz and a peak-to-peak value of 5 V is applied as an excitation; in order to avoid damage to the diode 2CZ106C, the PCB board uses a 100 Ω resistor as the diode series load. The local schematic diagram of the diode 2CZ106C reverse recovery waveform is shown in Figure 14. Among them, the divider of the 100 Ω resistor when the diode is forward bias is 1.20391 V, and the division of the 100 Ω resistor when the diode is reverse biased is −2.51967 V. Combined with the theoretical analysis, it can be calculated that the diode forward current $I_F = V_1/R_L$ = 0.012 A and the reverse current $I_F = V_1/R_L$ = 0.0252 A.

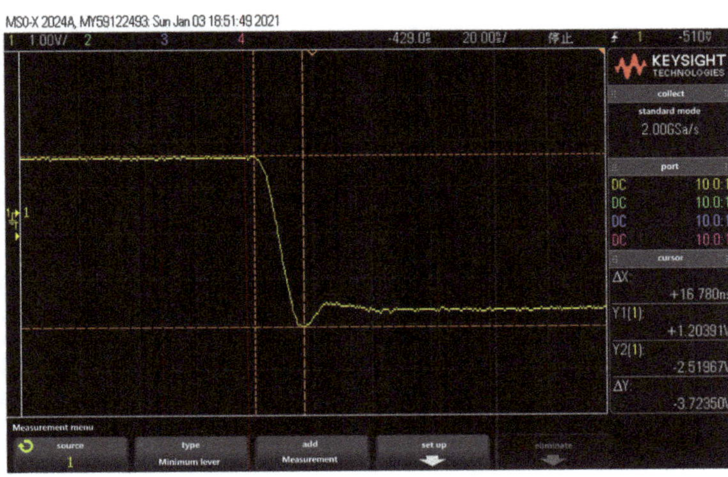

Figure 14. The reverse recovery waveform of 2CZ109C diode.

When the applied bias voltage is suddenly reversed, the diode current is also suddenly reversed, and the reverse current of t_s is constant over time, then the current slowly becomes smaller, and then over t_f time, the reverse current drops to $0.1I_R$. It is shown that when the reverse division of a 100 Ω resistor is about −0.251967 V, the diode is turned off in reverse. The diode 2CZ106C reverse recovery time can be read by the oscilloscope travel standard, as shown in Figure 15, and the diode 2CZ106C reverse recovery time t_r is 9.2918×10^{-7} s.

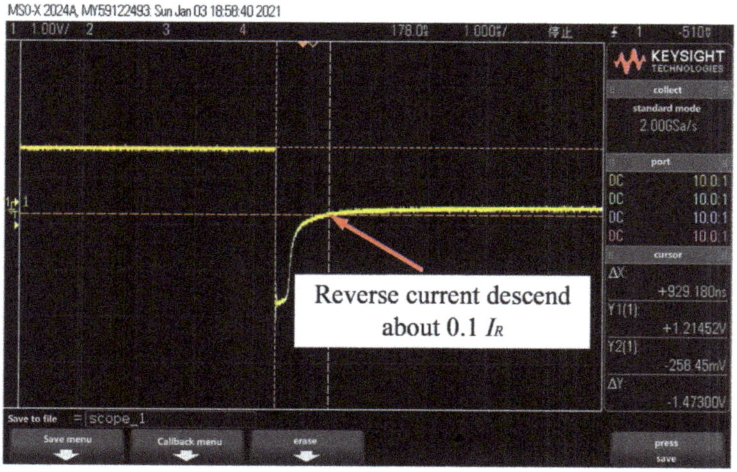

Figure 15. Test results of reverse recovery time t_r.

Therefore, the transit time characterizing the reverse recovery characteristic TT can be calculated by Equation (6).

$$TT = \tau_p = 0.4t_r / \ln\left[\frac{I_R + I_F}{I_R}\right] = 9.527 \times 10^{-7} S \quad (6)$$

3.4. Establishment of Diode Electromagnetic Compatibility SPICE Model

Through the test and data fitting method mentioned above, we obtain 6 of the 14 parameters necessary for diode modeling, and the other parameters can be obtained through the data sheet.

According to the electrical characteristics provided by the data sheet, the reverse breakdown voltage V_{BV} and reverse breakdown current I_{BV} of the diode 2CZ106C are determined to be 200 V and 10 μA, respectively.

The series resistance of the diode at high frequency is R_S and not pure resistance. To consider parasitic inductance and capacitance, the impedance frequency curve of the diode is tested using an impedance analyzer. The results show that the constant impedance of the diode in the frequency band below 10 MHz is about 90 Ω. To consider the influence of the parasitic parameters of the diode on the high-frequency operating signal, the series resistance is set to 90 Ω.

According to the semiconductor theory, the model parameters determined by the material process can be directly determined without optimization. Therefore, the reverse saturation current temperature index XTI, forbidden band width EG, flicker noise figure KF, flicker noise figure AF and forward biased barrier capacitance coefficient FC are all default values. In summary, the parameter values of the diode 2CZ106C model obtained through fitting calculation and analysis are shown in Table 4.

After obtaining the parameters necessary for diode modeling above, we can establish the electromagnetic compatibility model of the diode 2CZ106C according to SPICE grammar rules, as shown in the Figure 16.

```
*define sub-circuit
.subckt 2CZ106C 1 2
*the internal topology of diode
D1 1 2 DZ
*the value assignment of parameter base on .MODEL
.MODEL DZ D ( IS=4.378E-06 N=3.088 RS=90
+ BV=200 IBV=1.0E-5 CJO=5.150E-11 M=0.315
+ VJ=0.320 TT=9.517E-7 Af=1 KF=0 EG=1.11
+ FC=0.5 XTI=3)
*the termination statement of sub-circuit
.ends 2CZ106C
```

Figure 16. The EMC model of the diode 2CZ106C.

Table 4. Model parameters of diode 2CZ106C.

Number	Model Parameters	Value	Unit
1	reverse saturation current I_S	4.377×10^{-6}	A
2	emission coefficient N	3.088	-
3	series resistance R_S	90	Ω
4	reverse breakdown voltage V_{BV}	200	V
5	reverse breakdown current I_{BV}	1×10^{-5}	A
6	zero Bias Junction Capacitance C_{J0}	5.150×10^{-11}	F
7	gradient coefficient M	0.315	-
8	built-in potential V_J	0.320	V
9	transit time TT	9.517×10^{-7}	s
10	flicker noise index AF	1	-
11	flicker noise figure AF	0	-
12	band gap EG	1.11	eV
13	forward Biased Barrier Capacitance FC	0.5	-
14	reverse saturation current temperature index XTI	3	-

4. Model Simulation and Verification

4.1. Design of Model Verification Platform

The PCB board, shown in Figure 17, is designed and fabricated. To avoid introducing errors caused by too many other components, only 100 Ω resistors are used as loads to protect the diodes, which are shown in Figure 13; the model verification platform uses a signal generator to provide different frequency excitation signals and an oscilloscope to test the time-domain waveforms.

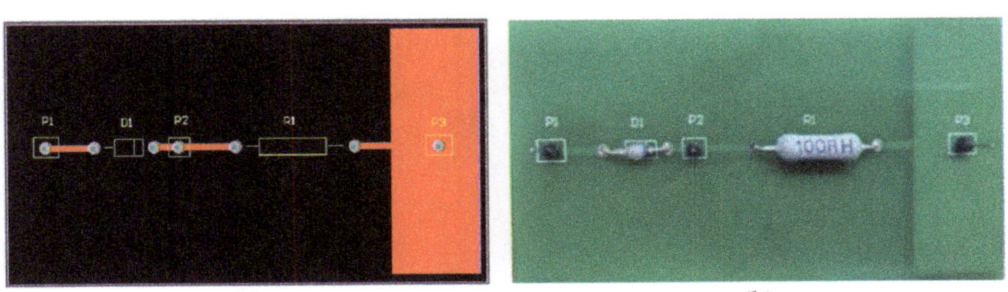

(**a**) (**b**)

Figure 17. PCB board for model verification (**a**) design drawing; (**b**) Processed circuit board.

To ensure that the input signals of the simulation and test are consistent, two channels are used to test and save the signals at the input end and load end of the PCB board, respectively, as the excitation signal for the subsequent simulation. Figure 18a–c shows the time-domain waveforms of the input end (one channel) and load end (two channels) of the PCB board when the signal frequencies are 10 kHz, 100 kHz and 1 MHz, respectively.

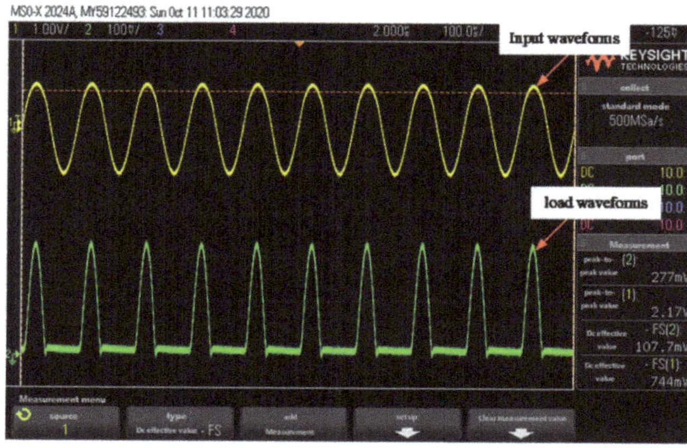

(**a**)

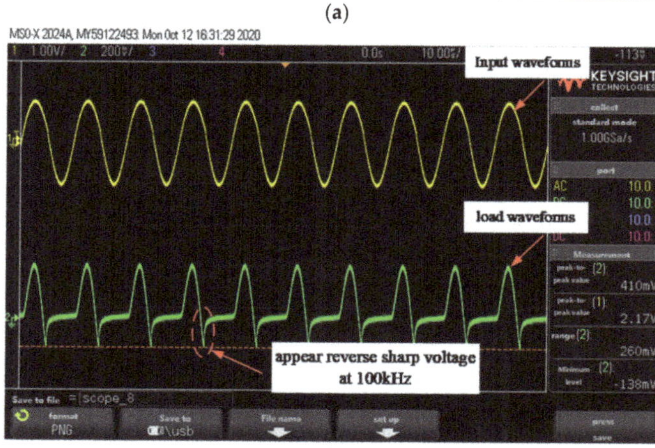

(**b**)

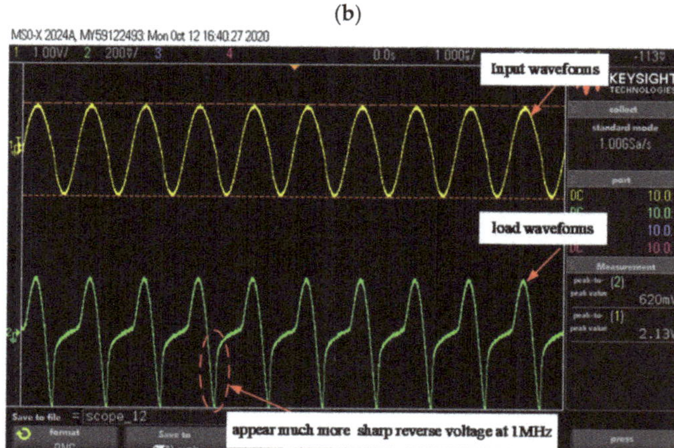

(**c**)

Figure 18. Experimental test results: (**a**) Input and load waveforms of 10 kHz excitation; (**b**) Input and load waveforms of 100 kHz excitation; (**c**) Input and load shape of 1 MHz excitation.

It can be seen that when the frequency of the excitation signal increases, the waveform of the load terminal is gradually distorted. The higher the frequency, the more irregular the waveform, which is caused by diode junction capacitance and reverse recovery characteristics.

From Figure 18, we can also see that when the excitation signal frequency is 10 kHz, the load resistor can achieve a half-wave output, indicating that the diode 2CZ106 can achieve good single guide passivity. When the excitation signal frequency is 100 kHz, the diode 2CZ106 does not fully achieve single passivity, although the shutdown time is short, but there is a large reverse voltage at the load resistor, with a peak value of about 0.15 V. When the excitation signal frequency reaches 1 MHz, the diode 2CZ106 loses almost single conductivity and there is a larger reverse voltage at the load resistor, with a peak of about 0.3 V.

4.2. Comparison with Simulation

The following is a comparison of simulation and test results. First, the diode verification PCB design file shown in Figure 17a is imported from the Altium Designer software into the CST PCB studio. In the import settings of CST, some unimportant components can be removed, thereby simplifying model complexity and improving simulation speed. After importing the PCB design file, we use the 2D TL Modeling function in CST PCB studio to generate the equivalent circuit model corresponding to the PCB board, as shown in Figure 19.

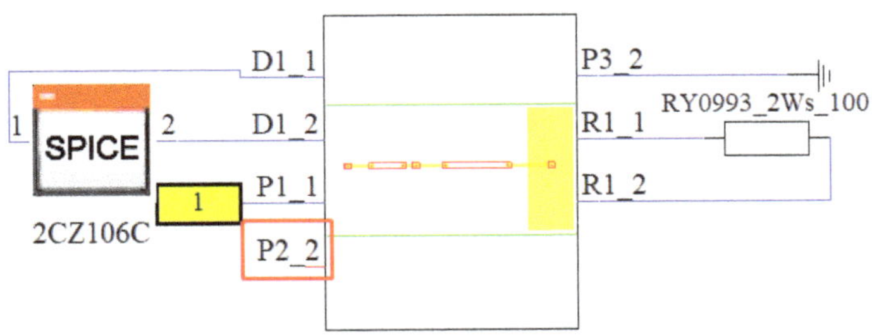

Figure 19. The complete simulation circuit established in CST.

The time domain signal at the input terminal of the PCB board shown in Figure 17 is used as the excitation source of the simulation model, and the time-domain waveform results of the simulation and measurement results are shown in Figure 20.

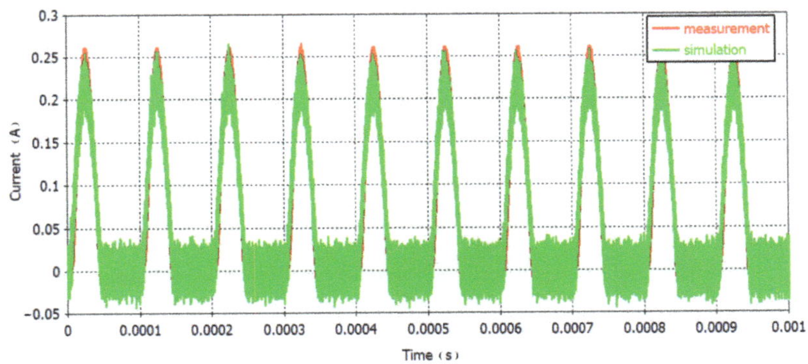

(a)

Figure 20. *Cont.*

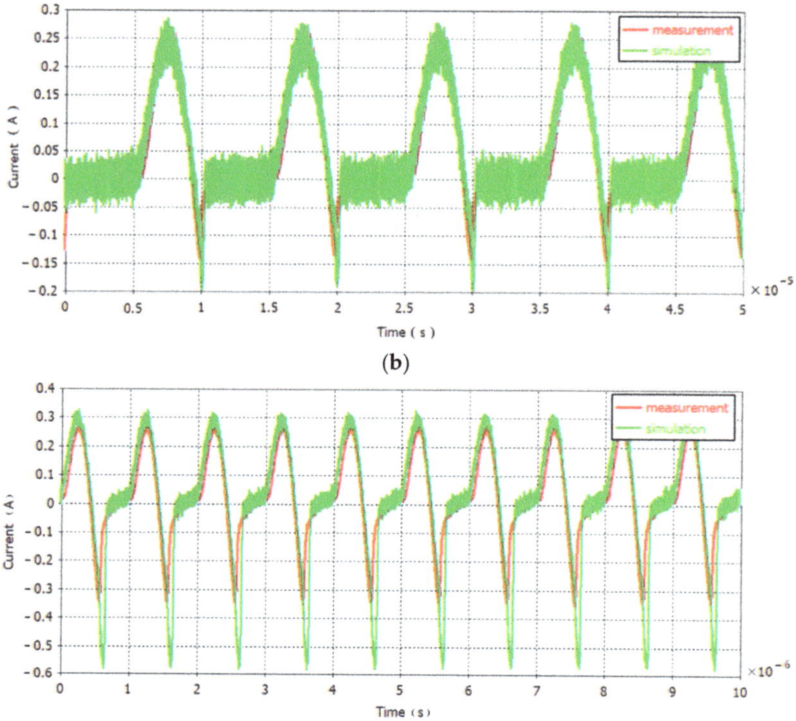

(**b**)

(**c**)

Figure 20. Comparison of measured and simulated in time-domain: (**a**) 10 kHz; (**b**) 100 kHz; (**c**) 1 MHz.

From the comparison of the above time-domain results, when the sinusoidal signal with a given frequency of 10 kHz is used as an excitation, the time-domain waveform results calculated by experimental testing and simulation are well matched. When the excitation signal frequency is 100 kHz and 1 MHz, there is a reverse voltage in the time-domain waveform of the measurement and simulation, and there is a slight error in their coincidence but the trend is basically the same.

This slight error may be due to the following reasons: First, the model parameter that reflects the reverse recovery characteristics of the diode is the transition time *TT*. From Equation (1), only a few parameters are considered to calculate the reverse recovery characteristics and these cannot fully reflect the reverse recovery process at high signal frequencies; Secondly, the data of the reverse recovery characteristics are mainly derived from measurements, and then the transition time *TT* is calculated by theoretical formulas. This process may introduce certain experimental errors, and the assumptions made in the derivation of theoretical formulas ignore the influence of some practical factors; Finally, the transition time *TT* has been modified during modeling, due to the irregularity of the internal carrier motion of the diode and the complexity of the internal electric field distribution, and there may still be some modeling errors after the approximate correction.

Considering that the measured time-domain waveform is not a standard sine wave, there are harmonic components in different frequency bands. By comparing the measured and simulated frequency components and the amplitude components of each frequency component, it can be verified whether the diode EMC model accurately reflects its related characteristics in the frequency domain. Therefore, the simulated and measured time-domain waveforms are converted into spectrum results by the fast Fourier transform function of CST, as shown in Figure 21.

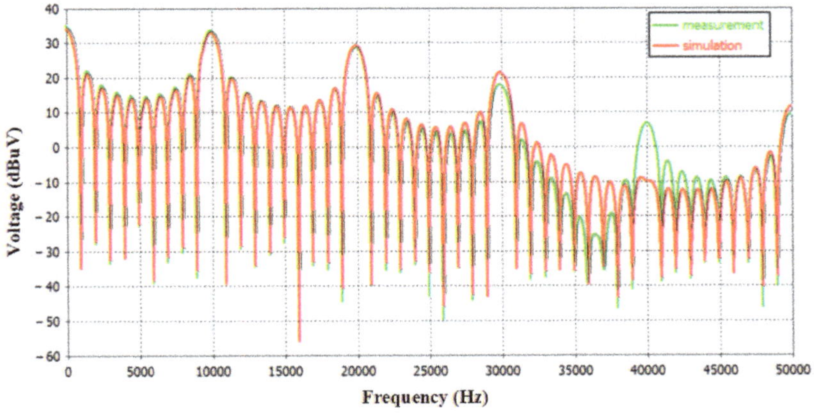

(**a**)

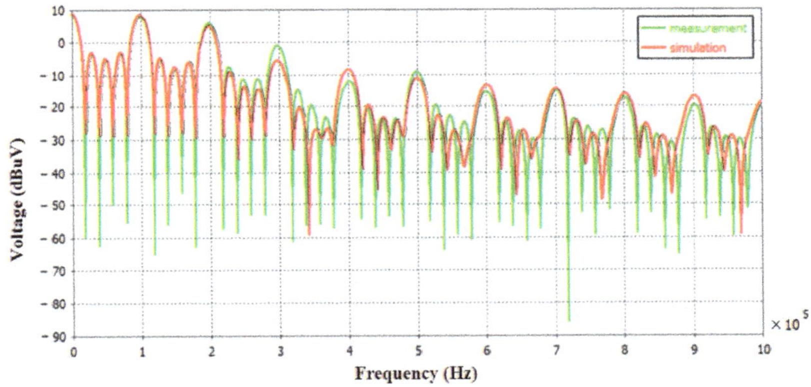

(**b**)

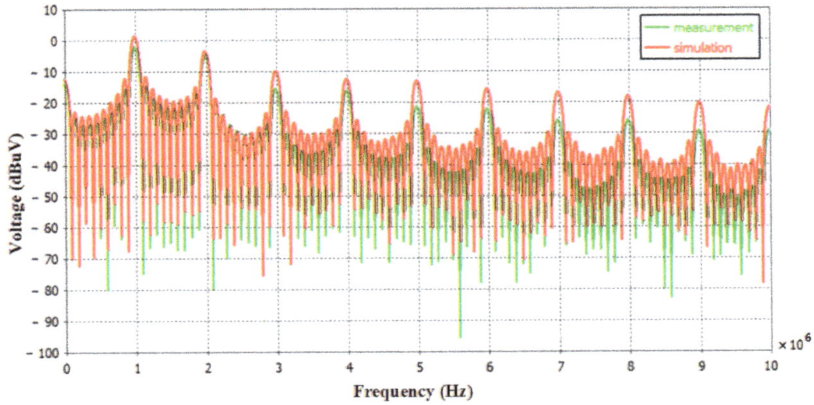

(**c**)

Figure 21. Frequency domain comparison of measured and simulated results: (**a**) 10 kHz; (**b**) 100 kHz; (**c**) 1 MHz.

As can be seen from Figure 21, the spectral results of the experiment and simulation are generally in good agreement, and the effective frequency band near the fundamental frequency of 10 kHz, 100 kHz and 1 MHz is almost identical. Some valuable conclusions are: (1) Comparing the 10 kHz spectrum results, the frequency domain results of experiments and simulation are completely consistent in the frequency band below 2 times the fundamental frequency (20 kHz), and there is a certain error in the noise harmonic higher than the 2 times fundamental frequency, but the overall difference is not large; (2) Compared with the 100 kHz spectrum results, it is known that the frequency domain results of experimental testing and simulation are completely consistent in the frequency band below 2 times the fundamental frequency (200 kHz), and the noise harmonic higher than the 2 × fundamental frequency has a small error; (3) Compared with the 1 MHz spectrum results, it is known that the frequency domain results of experimental testing and simulation calculation are completely consistent in the frequency band below 3 times the fundamental frequency (3 MHz), and there is a certain error in the noise harmonics above 3 times the fundamental frequency, but the overall difference is not much.

In order to quantitatively analyze the difference between simulation results and test results, the Feature Selective Validation (FSV) method is applied [23,24]. The FSV technology is an IEEE standard that shows the correlation between two sets of data according to a specific standard, which is widely used in the field of electromagnetic simulation [25,26]. The FSV technical evaluation index mainly includes: total amplitude difference measure (ADM), total characteristic difference measure (FDM) and total global difference measure (GDM). According to the ADM, FDM and GDM sizes, the degree of conformity between the data is described as excellent (0–0.1), very good (0.1–0.2), good (0.2–0.4), fair (0.4–0.8), poor (0.8–1.6) and very poor (>1.6). The simulation and measurement results in time-domain are analyzed based on FSV method, as shown in Table 5.

Table 5. The FSV evaluation results of the proposed method.

Waveform		**FSV**		
		ADM	**FDM**	**GDM**
Time domain	10 kHz	0.1137/very good	0.197/very good	0.2415/good
	100 kHz	0.1213/very good	0.1808/very good	0.233/good
	1 MHz	0.2776/good	0.4365/fail	0.5552/fail
Frequency domain	10 kHz	0.06636/excellent	0.03406/excellent	0.07973/excellent
	100 kHz	0.1121/very good	0.1368/very good	0.197/very good
	1 MHz	0.1118/very good	0.1463/very good	0.1974/very good

From Figures 20 and 21 and Table 5, we can comprehensively see that the improvement of the SPICE model of the diode can reflect its actual work, and can ensure that the test and simulation calculation results are completely consistent in a certain frequency band, although there are some errors in the higher harmonics, but the overall difference is not large; it can be concluded that this method has high accuracy.

5. Conclusions

In this study, a method for establishing an electromagnetic compatibility model of diodes based on physical equations is proposed. Compared with the CST software library model, the model in this study can truly reflect the gradual process of diode reverse cut-off and the peak voltage generated by dynamic characteristics. A comparison of the measured and simulated results from the time and frequency domains shows that the model in this study not only truly reflects the peak voltage generated by the electromagnetic interference characteristics of the diode, but also ensures that the measured and simulated components match within three times the fundamental frequency on the basis of accurately simulating the actual working conditions. The internal harmonic components and amplitudes of each component are completely consistent, and the model has a high electromagnetic compatibility simulation accuracy.

Future work could be carried out on the following aspects:

(1) This paper only considers the temperature effect of reverse current and reverse breakdown voltage; in fact, the diode is thermally sensitive as a semiconductor device, and

its forward current and forward conduction voltage will also be affected by temperature. Therefore, to more accurately consider the temperature characteristics of the diode, it is necessary to consider that the analog forward characteristics are affected by temperature.

(2) The influence of reverse recovery time is considered in the modeling of this paper, but with the advent of the 5G era, positive guidance may also interfere with high-frequency signals, which is worth studying in the electromagnetic simulation calculation of high-speed electronic systems.

Author Contributions: Conceptualization, methodology, validation, writing—original draft preparation, D.R. and G.X.; investigation, J.-Q.L.; data curation, Z.-Y.P. and X.Z.; writing—review and editing, D.R. and G.X.; supervision, P.-A.D.; funding acquisition, D.R. All authors have read and agreed to the published version of the manuscript.

Funding: This work was supported by the Major National Science and Technology Fund of China under Grant No. J2019-V-0012-0107, the Innovation and Development Fund of China Academy of Engineering Physics under Grant No. CX20210032, the Institute of Electronic Engineering Project Grant No. J21H-TS 02.

Data Availability Statement: Not applicable.

Conflicts of Interest: The authors declare no conflict of interest.

References

1. Ramdani, M.; Sicard, E.; Boyer, A.; Dhia, S.B.; Whalen, J.J.; Hubing, T.H.; Wada, O.; Coenen, M. The Electromagnetic compatibility of Integrated Circuits-Past, Present, and Future. *IEEE Trans. Electromagn. Compat.* **2009**, *51*, 78–100. [CrossRef]
2. Chew, W.C. Marriage of Computational Electromagnetics and Electromagnetic Compatibility. In Proceedings of the 2018 Joint IEEE EMC and APEMC symposium, Singapore, 14–17 May 2018.
3. Chew, W.C. Maxwell's Equations and Modern Electromagnetics after 150 Years and Role of Electromagnetics in EMC. In Proceedings of the IEEE APEMC Conference, Shenzhen, China, 18–21 May 2016.
4. Leila, F.; Anis, A. New simplified model for predicting conducted EMI in DC/DC converters. *Electr. Eng.* **2017**, *99*, 1087–1097.
5. Kurt, M.C.; Arnold, J.R.; Pieter, G.W. The Measurement and SPICE Modelling of Schottky Barrier Diodes Appropriate for Use as Bypass Diodes within Photovoltaic Modules. *Energies* **2022**, *15*, 4783. [CrossRef]
6. Ge, K.; Liping, J. *Analog and Digital Circuits*, 3rd ed.; Publishing House of Electronics Industry: Beijing, China, 2015; pp. 8–20.
7. Antognetti, P.; Giuseppe, M. *Semiconductor Device Modeling with SPICE*; McGraw Hill: New York, NY, USA, 1988; pp. 11–40.
8. Deveney, M. A temperature dependent SPICE macro-model for Zener and avalanche diodes. In Proceedings of the 34th Midwest Symposium on Circuits and Systems, Monterey, CA, USA, 14–17 May 1991; pp. 15–26.
9. Bley, M.; Filho, M.; Raizer, A. Modeling transient discharge suppressors. *IEEE Potentials* **2004**, *23*, 43–45.
10. Lepkowski, J.; Lepkowski, W. Evaluating TVS Protection Circuits with SPICE. *Power Electron. Technol.* **2006**, *32*, 44–49.
11. Piotrowski, R. An improved Spice2 Zener diode model for soft-region simulation capability. *IEEE Trans. Comput.-Aided Des. Integr. Circuits Syst.* **1988**, *7*, 1301–1303. [CrossRef]
12. Zhang, D.; Brown, R.; Growden, A.; Berger, P.R.; Droopad, R. A Nonlinear Circuit Simulation of Switching Process in Resonant-Tunneling Diodes. *IEEE Trans. Electron Devices* **2016**, *63*, 4993–4997. [CrossRef]
13. Tohlu, M.; Aoki, M.; Takashi, H.; Wada, O. Equivalent circuit model with nonlinear characteristics of Zener diode extracted from SPICE model for ESD simulation. In Proceedings of the International Workshop on the Electromagnetic Compatibility of Integrated Circuits, St. Petersburg, Russia, 4–8 July 2017; pp. 61–65.
14. Gan, Z. Simplified Zener Diode DC Spice Model. U.S. Patent 20160203250 A1, 14 July 2016.
15. Hitchcock, C.W.; Zhou, X.; Pande, G.; Ghandi, R.; Bolotnikov, A.; Chow, T.P. A Subcircuit SPICE Model for SiC Charge-Balance Schottky Diodes. *Mater. Sci. Forum* **2020**, *1004*, 945–952.
16. Yunyeong, C.; Miryeon, K.; Jisun, P.; Shin, H. Analysis of Organic Light-Emitting Diode SPICE Models with Constant or Voltage-Dependent Components. *J. Nanosci. Nanotechnol.* **2020**, *20*, 4773–4777.
17. Tao, L.; Le, X.; Yao, H.; Wu, H.; Yang, Y.; Wu, N.; Wei, F.; Shi, X.; Yang, X. A Novel Simulation Method for Analyzing Diode Electrical Characteristics Based on Neural Networks. *Electronics* **2021**, *10*, 2337. [CrossRef]
18. Yingjie, J.; Fei, X.; Yaoqing, D.; Yifei, L. Field-circuit Coupling Modeling and Multi-rate Electro-thermal co-simulation Analysis of Merged PiN Schottky SiC Diode. *Proc. CSEE* **2019**, *39*, 5585–5594.
19. Qiang, L.; Haijing, Z.; Fukai, X.; Zhenguo, Z. Research on Electromagnetic Interference of Switch Power Supply Based on PIN Diode Model. *J. Microw.* **2014**, *S2*, 29–32.
20. Chengkai, W. Research on Terahertz Frequency-Multiplied Technology Based on Schottky Diodes. Master's Thesis, University of Electronic Science and Technology of China, Chengdu, China, 2018; pp. 14–24.

21. Xingbi, C.; Yong, C. *Microelectronic Devices*, 3rd ed.; Publishing House of Electronics Industry: Beijing, China, 2018; pp. 70–74.
22. Enke, L.; Binsheng, Z.; Jinsheng, L. *Semiconductor Physics*, 7th ed.; Publishing House of Electronics Industry: Xi'an, China, 2008; pp. 181–185.
23. *IEEE Std. 1597.1-2008*; IEEE Standard for Validation of Computational Electromagnetics Computer Modeling and Simulations. IEEE: New York, NY, USA, 2008; pp. 1–41.
24. *IEEE Std. 1597.2-2010*; IEEE Recommended Practice for Validation of Computational Electromagnetics Computer Modeling and Simulations. IEEE: New York, NY, USA, 2011; pp. 1–124.
25. Alistair, D.; Gang, Z.; Slawomir, K.; Wang, L. Objective Selection of Minimum Acceptable Mesh Refinement for EMC Simulations. *IEEE Trans. Electromagn. Compat.* **2015**, *57*, 1266–1269.
26. Alistair, D.; Gang, Z.; Slawomir, K.; Orlandi, A.; Di Febo, D.; Wang, L.; Sasse, H. Comparison of Data with Multiple Degrees of Freedom Utilizing the Feature Selective Validation Method. *IEEE Trans. Electromagn. Compat.* **2016**, *58*, 784–791.

electronics

Article

Accurate Design of Microwave Filter Based on Surrogate Model-Assisted Evolutionary Algorithm

Yongliang Zhang [1], Xiaoli Wang [2], Yanxing Wang [2], Ningchaoran Yan [2], Linping Feng [3] and Lu Zhang [1,*]

[1] The College of Transportation, Inner Mongolia University, Hohhot 010021, China
[2] The School of Electronics and Information Engineering, Inner Mongolia University, Hohhot 010021, China
[3] The School of Electronics and Information Engineering, Xi'an Jiaotong University, Xi'an 710049, China
* Correspondence: imuzhanglu@imu.edu.cn

Abstract: Filter optimization problems involve time-consuming simulations and many variables in the design. These problems require a large amount of computation. This paper proposes an adaptive online updating 1D convolutional autoencoders (AOU-1D-CAE) surrogate model for solving this computationally expensive problem. In the optimization process, an adaptive update surrogate mapping between input variables and output objectives is constructed within the surrogate model AOU-1D-CAE framework. AOU-1D-CAE can replace electromagnetic (EM) simulation software for data collection, and select and automatically use the accumulated data as training samples to train the AOU-1D-CAE surrogate model. With more and more training samples, the learning ability of the surrogate model is also becoming stronger and stronger. The experimental results show that the data collection efficiency of AOU-1D-CAE is greatly improved, and the automatic update of the sample set improves the prediction performance of the surrogate model. In this paper, the optimization framework is AOU-1D-CAE-assisted particle swarm optimization (PSO), and the surrogate model assists PSO to find the global optimal solution. In the PSO stage, PSO automatically updates and saves the optimal solution, and takes the optimal solution of each stage as the initial solution of the next optimization stage to avoid falling into the local optimal solution. The optimization time is greatly saved and the optimization efficiency is improved. The continuous iteration of PSO also improves the prediction accuracy of the surrogate model. The efficiency of the proposed surrogate model is demonstrated by using two cavity filters as examples.

Keywords: microwave filter design; one-dimensional convolutional autoencoders; online updating; particle swarm optimization; surrogate model

Citation: Zhang, Y.; Wang, X.; Wang, Y.; Yan, N.; Feng, L.; Zhang, L. Accurate Design of Microwave Filter Based on Surrogate Model-Assisted Evolutionary Algorithm. *Electronics* **2022**, *11*, 3705. https://doi.org/10.3390/electronics11223705

Academic Editor: Alejandro Melcón Alvarez

Received: 13 September 2022
Accepted: 8 November 2022
Published: 12 November 2022

1. Introduction

Full-wave electromagnetic (EM) simulation-based optimization technology has become an essential tool for microwave filter design [1,2]. The purpose is to adjust the geometric parameter of the filter to ensure that the frequency response is within design specifications. Full-wave EM simulation is necessary for microwave device optimization [3], and its computational cost is very high. Optimizations on the basis of EM are also very time-consuming, as they usually require repeating EM simulations for various geometrical parameters as design parameters. Traditional EM optimization takes a long time to obtain the best solution to meet the specifications of the design. Therefore, obtaining the best solution in a relatively short period of time is a challenge.

The local optimization approach and the global optimization approach are two types of different methods, which are used in EM optimization, and applied to design and optimize filters. Several local optimization approaches were exploited for EM-based design closures, such as neighbor-sensitivity based on optimization [4,5], and space mapping based on optimization [6,7]. Nevertheless, because the problem is severely multi-patterned [8], local optimizations are inadequate and a global search needs to be performed. Various global optimization approaches like simulated annealing (SA) [9], particle swarm optimization (PSO) [10], and genetic algorithms (GA) [11] were successfully developed. The algorithms have a random nature which can assist in avoiding falling into a local optimum. The design of filters often has complex and multi-objective aspects. Reaching the optimal variable is related to the objective function under multiple objectives [12,13]. However, the shortcoming is the comparatively slow rate of convergence in these algorithms. In order to achieve a marked reduction in calculation costs, machine learning technologies are used to build surrogate models in place of computing costly EM simulations.

Artificial neural networks (ANN) [14], Kriging [15,16], Support vector machine (SVM) [17], and Gaussian processes (GP) [18] are some of the highly popular machine learning algorithms. They have many variants or combinations [19]. This is an efficient way of getting effective surrogate models. Such algorithms can specifically feature the unit cells and act as a replacement for full-wave EM tools with local regularity. Rapid alternative models are becoming increasingly important in the development and design automation of modern microwave device structures. There are various alternative assisted optimization methods for microwave devices to optimize the problem. The surrogate model assisted evolutionary algorithm (SAEA) is one of the most developed methods. The principle is to use a surrogate model instead of simulations. As the computation of the exact function evaluation is very expensive, the surrogate model uses much less computational overhead than the exact function evaluation of the simulation software. In [15], the forecasts of the various surrogate models are weighted to aid the evolutionary algorithm (EA), and the weights can be adaptively adapted according to the forecasts of the various surrogate models. Cai et al. introduced SAEA to the area of antenna design and synthesis optimization [20]. The surrogate model and the evolutionary algorithm are in a cooperative relationship in SAEA, in contrast with the surrogate model in the spatial mapping-based method. It is worth noting that with the widespread use of ANN in different fields, it is also gaining an interest in the field of microwave design. As a result, many surrogate models currently use ANN in the framework structure.

In recent years, as one of the most well-known machine learning methods [21,22], neural transfer functions (neuro-TF) [23] were developed, which are alternatives to the combination of neural networks and transfer functions and serve to parametrically model and optimize the design of microwave devices [24]. The use of ancillary feature frequencies derived from neural TF [25] is an agency-based model for EM optimization. The principle is to use feature frequencies to help with agent-based model EM optimization in order to avoid getting stuck in a local optimum in the optimization process. However, the traditional ANN model has too many layers, which makes the network more easily over-fitted. In addition to this, EA acts as a search engine in most SAEA, generating new points primarily for optimization, but as the EA generates more new data, there is no guarantee of the accuracy of the surrogate model and thus the feasibility of the EA generating new solutions. A one-dimensional convolutional autoencoders (1D-CAE) based surrogate model was first proposed in [26] to optimize the filter. However, the EM simulation software takes too long to collect training samples, is computationally intensive, and the overall optimization process tends to fall into local optimal solutions.

In this paper, a new approach, adaptive online updating 1D convolutional autoencoders-assisted particle swarm optimization (AOU-1D-CAE-APSO), is introduced which can solve this type of problem. The innovation of AOU-1D-CAE-APSO is the new SAEA framework, adaptive online updating 1D convolutional autoencoders (AOU-1D-CAE), consisting of a less complex ANN. The surrogate model constructs a non-linear relationship between the filter and the geometric parameters. After the training data has been collected by the EM simulation software to make the surrogate model achieve prediction accuracy, the surrogate model can replace the EM simulation software for training data collection. While accelerating the convergence of the optimization algorithm, the network parameters can be adaptively updated online to improve the prediction accuracy of the model embedded in the optimization algorithm.

The sections of this paper are listed below. Section 2 describes the fundamental technologies which make up the AOU-1D-CAE. Section 3 explains the specific work of AOU-1D-CAE-APSO. Section 4 describes two cavity filters using the AOU-1D-CAE-APSO method of optimization experiments and results. Section 5 provides a summarized conclusion.

2. Basic Theory

AOU-1D-CAE and PSO are the two key parts of this paper. AOU-1D-CAE is used as the surrogate model and the PSO algorithm is the search machine. A brief description of these techniques follows.

2.1. AOU-1D-CAE

Autoencoders (AE) is an excellent method for learning compression and distributed feature representation from a given dataset. The AE comprises an input layer, a hidden layer, and an output layer. A coding process between the input layer and the hidden layer is used to obtain an encoded representation through the coding operation on the input data. Between the hidden layer and the output layer, there is a decoding process, which obtains the reconstructed input data by decoding operations on the hidden layer. Figure 1 shows the basic framework of AE. The encoding process can be expressed as:

$$y = f(Wx + b) \quad (1)$$

where x is the input signal, y is the hidden layer, and f is a nonlinear function. W is the weight matrix. b is the bias matrix. The formula for the decoding process is:

$$\hat{x} = f'(W'y + b') \quad (2)$$

where $\hat{x}$ is the decoded output, which is the actual output of the network, and the predicted value of the input layer x, f' is a nonlinear function. W' is the weight matrix. b' is the bias matrix.

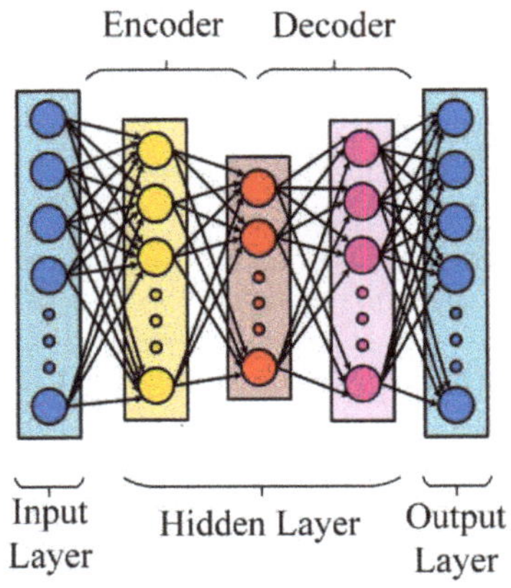

Figure 1. AE framework.

In this paper, the inputs to the encoding process are the real and imaginary parts of the S-parameters, and the outputs are the geometric parameters. The output of the encoding process is used as the input to the decoding process, and the reconstructed S-parameters are used as the output of the decoding process.

The purpose of the AE is to reconstruct the input data, which means that the AE is allowed to learn a function of $x = \hat{x}$. This allows the network to learn different representations of the input data that characterize the data. The AE, in order to reproduce the source input, has to capture the parts that can represent the important features of the source data. The disadvantage of the traditional AE model is the difficulty in extracting features of the input data. Thus, the AOU-1D-CAE model in this paper combines the 1dimensional convolutional neural network (1D-CNN) with the AE, which means the encoding and decoding process is performed by 1D-CNN. Figure 2 shows the framework of a 1D-CNN.

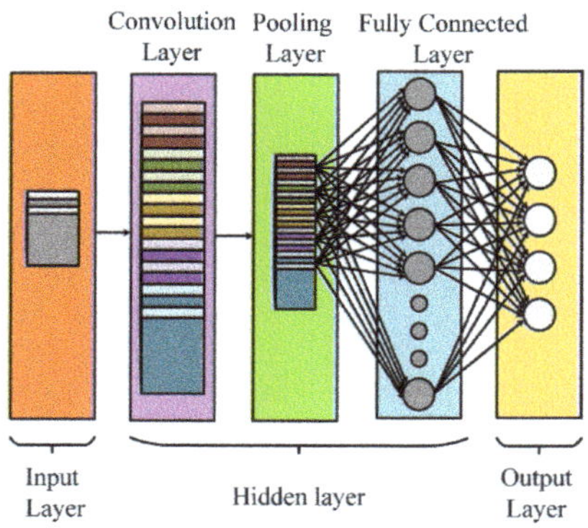

Figure 2. General CNN structure.

CNN is a feedforward neural network with a deep network structure. The principle is to extract features from the input data by means of multiple filters. The basic layer types in this paper model are the convolution layer, pooling layer, and fully connected layer. The Convolutional (Conv) layer is the basic structural unit, and the Conv kernel parameters are shared globally, reducing parametric numbers. The number of feature layers can be controlled using the size of the feature map in the Conv layer to learn more and more physical features. The calculation formula of the Conv layers is as follows:

$$y_j^{l+1} = f(\sum_{i=1}^{M} x_j^i \otimes k_{ij}^l + b_1^l) \quad (3)$$

where y_j^{l+1} is the input of j-th neuron at layer l + 1; $f(\cdot)$ is activation function; M is the feature map; x_i^l which is the output of neuron *j-th* of layer l; $\otimes$ means convolution operation; k_{ij}^l is the convolution kernel of *i-th* neurons in layer l with *j-th* neurons in layer l + *1*; b_i^l is the bias.

The pooling layer reduces the amount of data and network overfitting during the solving process. What is used is to extract the maximum value of the pooling window as the output result and then keep sliding the window to reduce the size of the input data, thus reducing redundant information. The calculation process of maximum pooling is as follows:

$$p_j^{l+1}(i) = \max_{(i-1)W+1 \le t \le iW} \left\{ a_j^l(t) \right\} \quad (4)$$

where $p_j^{l+1}(i)$ is the corresponding value neurons in layer l + *1* of the pooling operation; W is the width of the pooling region.

A fully connected layer expands the input to the previous layer with a column vector, with an all-connected layer, it acts as a classifier in CNN. The fully connected layer is computed as follows:

$$z_j^{l+1}(i) = \sum_{i=1}^{n} w_{ij}^l p_j^l + b_j^l \quad (5)$$

where z_j^{l+1} is the output of *i-th* neurons of layer l + *1*; w_{ij}^l is the weights of *i-th* neurons in layer l with *j-th* neurons in layer l + 1; P_j^l represents the corresponding value of the neuron in layer l of the pooling operation; b is the bias.

2.2. PSO

In the AOU-1D-CAE-APSO algorithm, PSO is used as a search engine. PSO is an evolutionary algorithm that simulates the feeding behavior of a flock of natural species of birds. When searching for food in an area, birds are not sure how far away their current position is from the food, so they need to search the area around the nearest bird to the food. The particle is the simulation of the bird in the algorithm. Each particle has two attributes: velocity and position. Particle i has two properties, velocity, and position. At time g, the velocity and position of the particle are denoted by the subscripts $v_{i,g}$ and $x_{i,g}$. The velocities show the steps of motion and the positions represent the directions. Each particle finds an individual optimal solution, and the global optimal solution is obtained by comparing the individual optimal solutions found for all particles. Multiple iterations keep the velocity and position updated until the termination condition is met when the loop is exited and the iteration ends.

The velocity and position of each individual (particle) are renewed by the following equation:

$$v_{i,g+1} = v_{i,g} + c_1 \times rand_{i,1} \times (x_{i,pest} - x_{i,g}) + c_2 \times rand_{i,2} \times (x_{gpest} - x_{i,g}) \quad (6)$$

$$x_{i,g+1} = x_{i,g} + v_{i,g+1} \quad (7)$$

$x_{i,g} = (x_1, x_2, \ldots, x_d)$ represents the position of the *i-th* individual in the d-dimensional search space at the current PSO iteration to the *g-th* generation, $v_{i,g} = (v_1, v_2, \ldots, v_d)$ represents the speed of the *i-th* individual in the d-dimensional search space, $x_{i,pbest}$ represents the historical optimal position of the *i-th* individual in the d-dimensional search space, x_{gbest} represents the optimal position of all individuals of the entire population in the d-dimensional search space during the optimization process. c_1 and c_2 represent the learning rates for learning from the *i-th* individual historical optimum and the global optimum, respectively, $rand_{i,1}$ and $rand_{i,2}$ is the random number between [0, 1]. The flow of the PSO algorithm is shown in Figure 3.

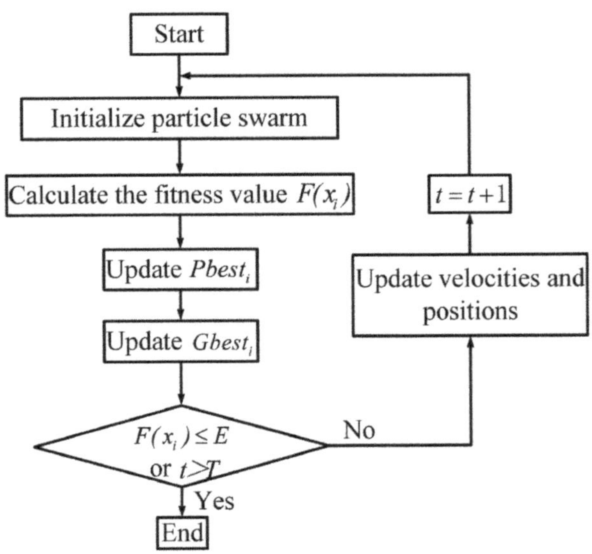

Figure 3. The flow of PSO.

3. AOU-1D-CAE-APSO Algorithm

3.1. Algorithm Framework

The AOU-1D-CAE-APSO framework is shown in Figure 4, which works as follows:

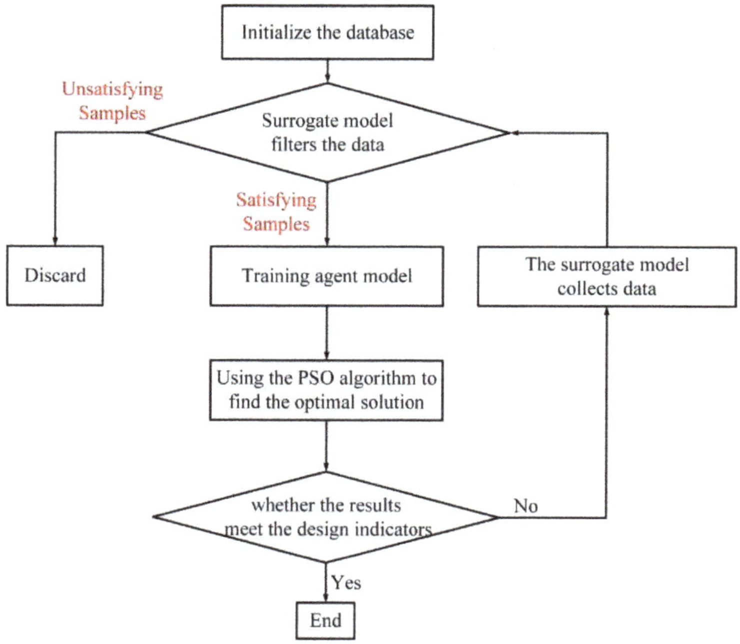

Figure 4. Framework of AOU-1D-CAE-APSO.

Step1: Python and High-Frequency Structure Simulator (HFSS) co-simulation to form the initial database;
Step2: Select the samples in the database, and if the preset conditions are met (e.g., the samples in the database meet the satisfaction criteria), then keep the data sample; otherwise delete the sample;
Step3: The database samples are trained to make the surrogate model with certain accuracy and prediction function;
Step4: Add an optimization algorithm that acts as a search engine for the overall optimization process. Embed the surrogate model into the optimization algorithm and use the update iterations of the algorithm to find the optimal solution;
Step5: Determine if the design specifications are met, and if the preset stopping criteria are met (e.g., the reflection coefficient in the optimized pass-band range is reduced to below −20 dB), output the best solution from the optimization program; otherwise go to Step 6;
Step6: The surrogate model captures data to form a second generation database;
Step7: Select the samples in the database, if the preset conditions are met (e.g., the samples in the database meet the satisfaction criteria), then keep the data sample; otherwise delete the sample;
Step8: Combine the initial database samples and the second generation database samples to train the agent model, so that the surrogate model has a certain accuracy and a prediction function;
Step9: The optimization algorithm performs the optimization and comes up with the optimal solution;
Step10: Determine if the design specifications are met, if the preset stopping criteria are met (e.g., the reflection coefficient in the optimized pass-band range is reduced to below −20 dB), then output the best solution from the optimization program; otherwise go to Step 6.

In the data collection stage, the collected real data content is the geometric parameters and their corresponding S parameters. The input of the autoencoder is the S-parameters of the real data, and the input of the PSO is the geometrical parameters of the real data.

The adaptive online update in the surrogate model structure has two innovations: (1) Replaces EM simulation software to reacquire data. A significant saving in filter design time and increased efficiency. (2) When the network prediction is not accurate, the surrogate model automatically collects data near the predicted value and retrains the network to make the network more accurate and improve the prediction performance.

3.2. Parameter Settings of AOU-1D-CAE-APSO

There are several parameter settings in the AOU-1D-CAE-APSO, some experimental principles are presented in this paper.

Data Pre-processing. In [27], it is stated that the neural network model accepts the geometry and frequency variables of the filter as input. The values of the geometric and physical frequency variables can change by many orders of magnitude. To address the problem that for *s*-shaped neurons, too large an input can lead to saturation, when the derivative of the *s*-shaped function is very close to zero, slowing down the learning process of the ANN model, data pre-processing is required. Within the range of values of the input variable x $[x_{min}, x_{max}]$, it can be linearly scaled to $[-1,1]$:

$$\overline{x} = 2\frac{x - x_{min}}{x_{max} - x_{min}} - 1 \quad (8)$$

Equation (8) shows the linear mapping between the physical value x and the ANN model input value $\overline{x}$. The mapping calculates the physical value x to obtain the optimized neural network model input value x:

$$x = \frac{(x_{max} - x_{min})}{2}(\overline{x} + 1) + x_{min} \quad (9)$$

Parameter setting in the surrogate model. The number of training data is decided by the trade-off between the surrogate model quality and the surrogate model training time. In this paper, the optimization process is intelligent. When the accuracy of the surrogate model is not high enough, training sample data can be added automatically to train the surrogate model. The detailed internal structure of the surrogate model is shown in Figure 5, including the layers, size/Nodes, Stride, and activation functions for each layer. The network structure of the surrogate model is shown in Figure 6.

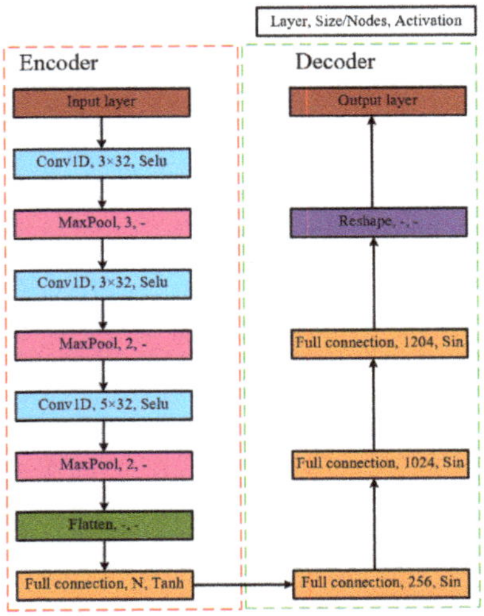

Figure 5. Detailed architecture of 1D−CAE in AOU−1D−CAE−APSO.

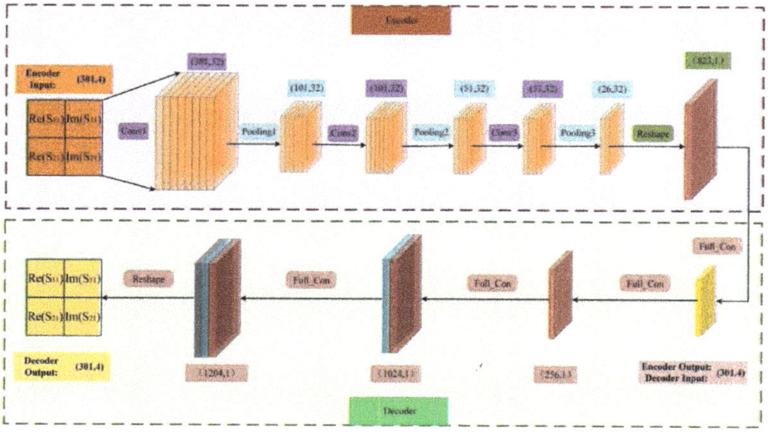

Figure 6. Network structure of the AOU-1D-CAE.

The loss function of the surrogate model is cited from the literature [26]: a combination of two loss functions, reconstruction loss (L_r) and prediction loss (L_p). The combined loss function is used to train the entire network. In the AOU-1D-CAE surrogate model, the real and imaginary parts of the input data S_{11} and S_{21} are encoded as geometric parameters. The decoder reconstructs the real and imaginary parts of S_{11} and S_{21}. Reconstruction losses were used to evaluate the reconstruction performance of the AOU-1D-CAE models. The reconstruction loss is expressed as follows:

$$L_r = \frac{1}{N}\sum_{i=1}^{N}|S_t - S_p| \tag{10}$$

where N is batch size. S_t is the simulated S-parameters, S_p is the reconstructed S-parameters of the 1D-CAE model. The prediction loss is defined as the mean square error between the predicted value of the encoder and the actual value of the geometric parameters:

$$L_p = \frac{1}{N}\sum_{i=1}^{N}|H_t - H_p| \tag{11}$$

where N is batch size. H_t is the actual value of the geometric parameters, H_p is the prediction value of the encoder. Finally, the overall loss function is the sum of L_r and L_p, which is defined as follows:

$$L = L_r + \lambda L_p \tag{12}$$

where λ is a regularization parameter. The loss function of this network is optimized by using the Adam optimizer.

The parameters in the search framework: in PSO, the position of the particles is the geometric parameter to be optimized in the filter. The search dimension is the number of geometric parameters. The fitness function, cost, takes points evenly across the range of the pass-band to be optimized by S_{11}, takes the distance between each point and the target value, and sums these values. PSO is responsible for searching for the position of the particle corresponding to the minimum value of the cost function.

4. Experiment and Results

4.1. Fourth-Order Cavity Filter

In the first example, consider the optimization of a fourth-order cavity filter [28] to demonstrate the proposed optimization technique. Figure 7 illustrates the structure of a fourth-order cavity filter. The fourth-order cavity filter is centered at 2.0693 GHz and the bandwidth is 110 MHz. The design specifications are defined as $|S_{11}| \leq -20$ dB in the frequency range from 2.0143 GHz to 2.1243 GHz. Each EM simulation costs about 3–4 min on average. The fourth-order cavity filter needed to optimize the geometric parameters are $X = [p_1, p_2, p_3, p_4]^T$. The optimization objective is the minimization of the fitness function, cost in Equation (13), to satisfy the design specifications. In PSO, the fitness function is:

$$Cost = \sum_{n=1}^{6} d_n - (-20) \times 6 \tag{13}$$

where d_i is the distance between the six points and the optimization target within the optimization pass-band on S_{11}, after taking six points at equal intervals. *Cost* is the fitness function of PSO, which seeks to find the minimum. The internal process of the AOU-1D-CAE-APSO is shown in Figure 8.

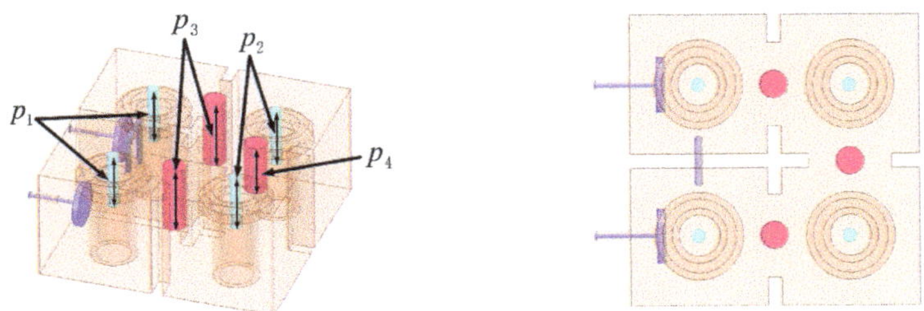

(**a**) (**b**)

Figure 7. Fourth-order cavity filter. (**a**) Side view. (**b**) Top view.

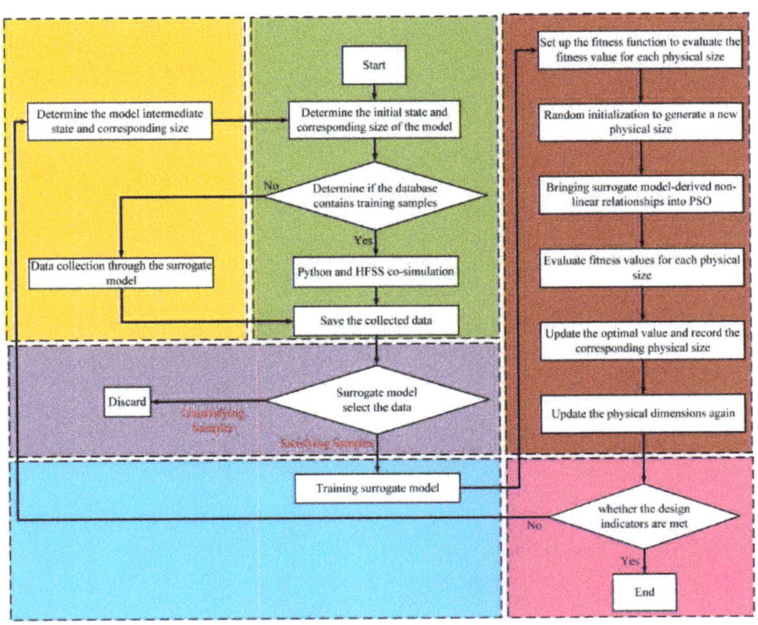

Figure 8. Internal process of the AOU-1D-CAE-APSO.

During the design of the microwave filter, the initial state of the filter is not ideal, use this stage as stage 1. The S_{11} of the filter before optimization is −5.3 dB, corresponding to the geometric parameters taken as $p_1 = p_2 = p_3 = p_4 = 10$ mm. The response results of the

PSO algorithm iteration are shown in Figure 9. The initial state is shown as the first stage in Figure 9.

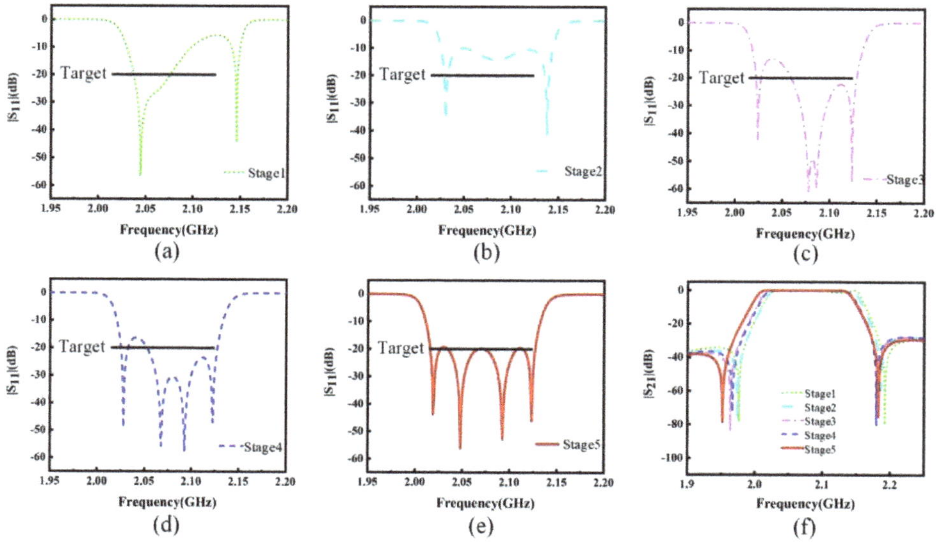

Figure 9. Fourth−order cavity filter HFSS verification of the optimized response in PSO iterations. (**a–e**) Reflection coefficient S_{11}. (**f**) Transmission coefficient S_{21}.

Firstly, add training samples. This paper proposes an optimized design method that only uses a co-simulation of python and HFSS for the first data collection. More data is obtained by disturbing the geometrical parameters, as in Equation (14),

$$p_n' = p_n \pm p_n \times 0.5\% \ (n = 1, 2, 3, 4) \tag{14}$$

the new geometrical parameters obtained are input to HFSS for simulation. For example, $p_1 = 10$ mm, $p_1' = p_1 \pm p_1 \times 0.5\%$, get a new set of geometric parameters $H' = [p_1', p_2, p_3, p_4]^T$, the remaining geometric parameters get new data according to this disturbing index, consist of new geometric parameters, then input into HFSS for simulation, get the sample data. The HFSS outputs are $\text{Re}(S_{11})$, $\text{Im}(S_{11})$, $\text{Re}(S_{21})$, $\text{Im}(S_{21})$. The sweep range is 1.85–2.2886 GHz, and each curve consists of 301 points. The resulting data are saved as npy files, the content of each npy file containing the geometric parameters, and the corresponding four S-parameter curves. The first data collection in this experiment was obtained from HFSS simulations and 300 data samples were collected over 930 min. Save as npy file. The samples are classified into 240 training data sets and 60 test data sets. Approximately 20 pieces of data were collected on average per hour.

Next step, select data. A selection range is usually defined. In this experiment, the initial state S_{11} is −5.3 dB, and the pass-band range is from 2.0143 to 2.1243 GHz. The optimization target is $S_{11} \leq -20$ dB. Set the selection range to [−10 dB, −5.3 dB], the data not in this range will be deleted, and the satisfactory data will be kept. 297 of these files met the conditions. 3 files were deleted.

Then start building the AOU-1D-CAE surrogate model. AOU-1D-CAE is a combination of AE and CNN. Generally divided into two parts, encoder and decoder. Train the surrogate model. The collected data samples were put into the surrogate model and the neural network was trained. $\text{Re}(S_{11})$, $\text{Im}(S_{11})$, $\text{Re}(S_{21})$, $\text{Im}(S_{21})$ from a real sample is used as input to the surrogate model, with each S-parameter consisting of 301 points. First, an encoding operation is performed. In the encoder, the input matrix is 301 × 4. After three mutually alternating convolution and pooling layers, the matrix becomes 26 × 32. After flattening the matrix becomes a vector, and then after a full-connected layer, prediction is performed. The output of the encoder is the corresponding geometric parameter obtained by the prediction of the S-parameter. The output of the encoder is then used as the input to the decoder for the decoding operation. After three fully connected layers in the decoder, the network is then reshaped to a matrix 301 × 4. The loss function of this network is Equation (12). The neural network learns the non-linear relationship between the geometrical parameters and the corresponding S-parameters and is able to predict the physical performance corresponding to any set of geometrical parameters with a loss function that meets the criteria. After building the surrogate model, the surrogate model was trained on NVIDIA Geforce GTX 1060 6 GB using the GPU environment. Experiments show that it takes 7 min to train the network using the GPU environment and 27 min using the CPU environment. Each time the surrogate model is trained, the prediction accuracy of the network is checked. A random set of geometric parameters X is chosen to be placed into the AOU-1D-CAE and the network outputs a set of S-parameters. Figure 10a illustrates the training and testing performance of the network. It can be seen that the model has a faster convergence rate and higher accuracy as the number of training sessions increases. Figure 10b presents a contrast between the HFSS simulation results and the predicted results from the AOU-1D-CAE model. It can be seen from Figure 10 that the results match exactly.

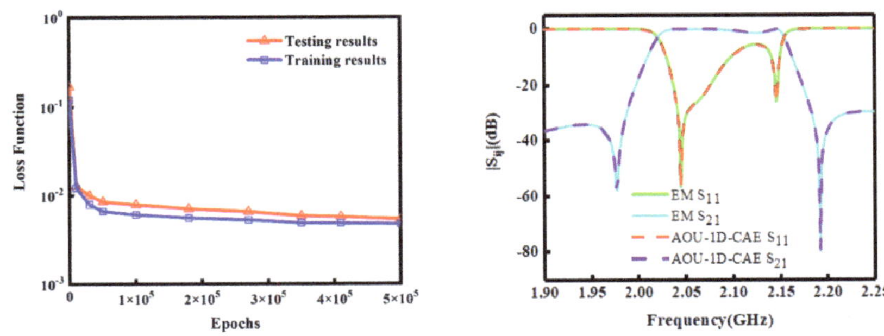

Figure 10. (**a**) AOU−1D−CAE neural network training and testing result. (**b**) Fourth−order cavity filter HFSS simulation results with surrogate model network prediction results.

The PSO were implemented using the open source and well-developed Python codes. The fitness function is the key to the PSO algorithm's important basis for evaluating whether the computed solution satisfies the design objective. The bandwidth to be optimized is intercepted, that is, S_{11} whose frequency range is between 2.0143 GHz and 2.1243 GHz is intercepted. Take 6 points (d_1, d_2, d_3, d_4, d_5, d_6) at equal intervals on the intercepted S_{11}, take the distance between these 6 points and $y = -20$ dB, and the sum is the fitness function of PSO. The fitness function is Equation (13), and the goal is to find the minimum value.

In this experiment, the particle dimension is 4 (the number of geometric parameters is 4). The number of particles is 100. The maximum number of iterations is 100. The acceleration factor $c_1 = c_2 = 2$. Once initialization is complete, a new set of geometric parameters is generated at random, bringing surrogate model-derived non-linear relationships into PSO. The resulting geometric parameters are fed into the surrogate model and the S-parameters are obtained by prediction. Calculate the fitness function for each particle. Update x_{gbest} and $x_{i,pbest}$.

The position of the x_{gbest} is at this point the optimal size parameter for the filter. After PSO iterations, the optimal particle in the population is derived as follows: P_1 = 9.7275 mm, P_2 = 10.1420 mm, P_3 = 9.6737 mm, P_4 = 9.6804 mm. The corresponding values are put into the surrogate model for prediction and a set of −9.8 dB reflection coefficients is obtained. Determine if the responses meet the design specifications. If the result does not meet, go to the next step.

As shown in Figure 9b, the model has been optimized from stage 1 to stage 2, with a decrease in S_{11} in the optimized pass-band. The reflection coefficient corresponding to stage 2 is −9.8 dB. At this point, the geometric parameters corresponding to stage 2 are used as the initial state and the training sample data are collected. The second data acquisition was performed using a surrogate model instead of HFSS simulation software. AOU-1D-CAE replaces HFSS for data acquisition, the same perturbation is carried out using Equation (14) to obtain more new geometric parameters. The new geometric parameters are put into the surrogate model as input and predicted by the surrogate model to produce *S*-parameters from which new training sample data can be composed. The second data collection in this experiment was obtained from AOU-1D-CAE simulations and 300 data samples were collected over 1 min. Save as npy file. The contents of the npy file are the same as the first time training data was collected.

Data samples in the range [−15, −9.8] dB are selected, and data not in this range are deleted. The AOU-1D-CAE obtained 300 npy files and took 1 min. 297 of these files met the conditions. 3 files were deleted.

The 297 data samples from the first collection were fused with the 297 data samples from the second collection and put into the surrogate model for training.

Bringing surrogate model-derived non-linear relationships into PSO. After PSO iterations, the optimal particle in the population is derived.

The position corresponding to the optimal particle, the geometric parameter *H*, is put into the surrogate model for prediction, and a set of reflection coefficients is obtained. At this point the reflection coefficient is 13 dB, corresponding to stage 3 of Figure 9c, which does not meet the design specification. Confirming this set of geometric parameters *H* as the initial state, the data around this set *H* is collected using the surrogate model with reference to the previous perturbation indicators.

Repeat the optimization process, meeting the design specifications when proceeding to stage 5. Exit the cycle. Figure 11 shows the details of each stage of the AOU-1D-CAE.

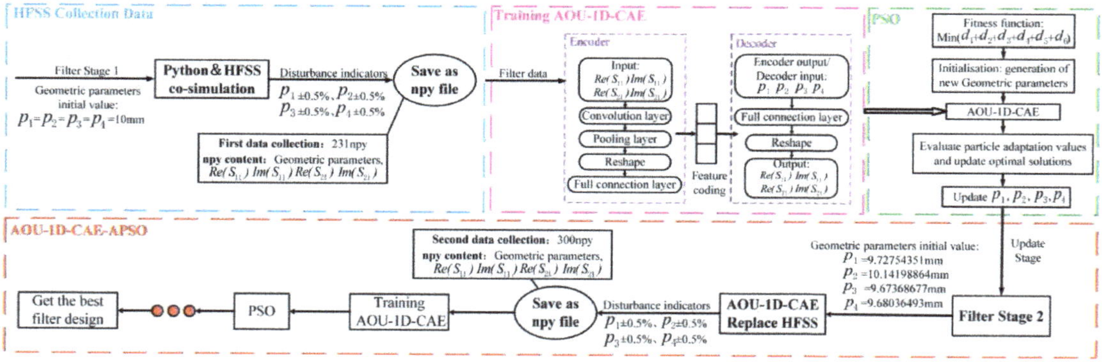

Figure 11. Details of the AOU-1D-CAE optimization.

AOU-1D-CAE-APSO obtains greatly increased speed improvement compared with standard full-wave simulation for this example, making the unbearable time to be very practical and obtaining an even better result. As shown in Table 1, comparing the two ways of collecting training data samples, some observations can be made. Optimizing filters with traditional full-wave simulations can take hundreds of minutes to calculate, whereas the optimization method proposed in this paper reduces the time greatly, and the surrogate model is able to collect 300 sets of data samples in 1 min, compared to 930 min for HFSS. The comparison of time, as well as the computational cost, is very obvious. Table 2 gives a comparison of the time taken to optimize the filter by the traditional optimization approach [26] and by the two surrogate model-assisted PSOs.

Table 1. Fourth-order cavity filter optimization process.

Stage	Initial State (dB)	Data Collection	Time	Satisfaction Data	Reflection Coefficient Distributed Area (dB)
1	−5.3	300	930 min	297	[−9.8, −5.3]
2	−9.8	300	1 min	297 + 297	[−13, −9.8]
3	−13	300	1 min	295 + 297 + 297	[−16.2, −13]
4	−16.2	300	1 min	286 + 295 + 297 + 297	[−20, −16.2]
5	−20	—	—	—	—

Table 2. Two surrogate model-assisted particle swarm optimization algorithms.

Step	Time	
	Traditional SEAE	AOU-1D-CAE-APSO
Initialize the database	930 min	930 min
Train the surrogate model	27 min	7 min
PSO	1 min	1 min
Adding training samples	930 min	1 min

4.2. Eighth-Order Cavity Filter

The first experiment optimized four parameters of the filter, this experiment will optimize eight parameters and prove whether the algorithm works.

In this section, experiments were carried out with an eighth-order cavity filter [29], the model of which is illustrated in Figure 12. The metrics are as follows, with a center frequency of 2.0693 GHz and a bandwidth of 110 MHz, and the geometrical parameters to be optimized are $H = [p_1, p_2, p_3, p_4, p_5, p_6, p_7, p_8]^T$. Each EM simulation costs about 12–15 min on average. The design specification for this example is defined as $|S_{11}| \leq -20$ dB in the frequency range of 2.0143–2.1243 GHz. The first time data was collected by HFSS, 485 data were collected in a total of 52.13 h. The data were collected simultaneously by two servers. The optimized pass-band range is 2.0143–2.1243 GHz. The experimental procedure was the same as in Section 4.1. The same parameters and fitness functions are used in PSO. The optimization objective is the minimization of the fitness function, cost in Equation (13), to satisfy the design specifications. Figure 13 presents a contrast between the HFSS simulation results and the predicted results from the AOU-1D-CAE model. Figure 14 illustrates the optimization process, the optimization process goes through four cycles and five stages. The S_{11} change from stage 1 to stage 5 corresponds to Figure 14a–e. Figure 14f shows the variation of S_{21}. The value of stage five approximates meeting the design requirements. After four cycles, PSO finds the optimal solution. After EM simulation, the design purpose is approximation satisfied. Stage 5 as the final result of the reflection coefficient, differs from the ideal result by −1.4 dB, which is within the acceptable margin of error. Table 2 shows the eighth-order cavity filter optimization process and the data collected. HFSS took 52 h to collect around four hundred data samples, whereas the agent model can give thousands of training data samples in just 4–6 min, so it can be seen that AOU-1D-CAE-APSO saves a lot of computational and time costs.

This experiment illustrates the feasibility of the proposed algorithm. The content of Table 3 is the eighth-order cavity filter optimization process. Table 3 reflects that the advantages of this algorithm for efficient and fast filter design are more evident when more parameters are optimized.

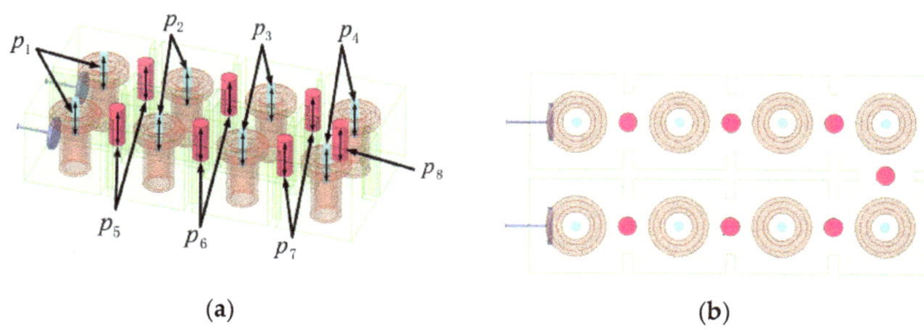

Figure 12. Eighth-order cavity filter. (**a**) Side view. (**b**) Top view.

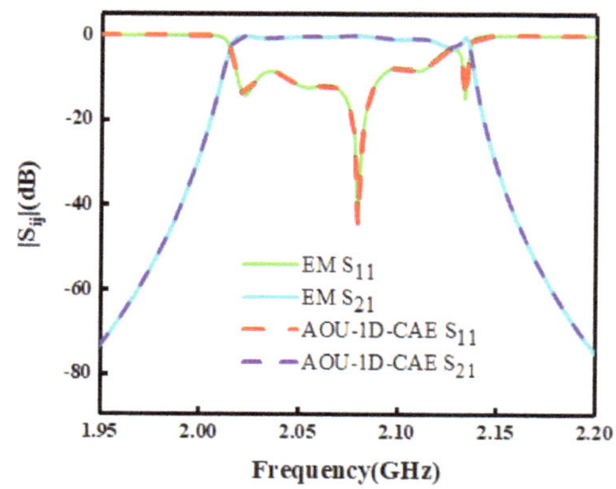

Figure 13. Eighth−order cavity filter HFSS simulation results with proxy model network prediction results.

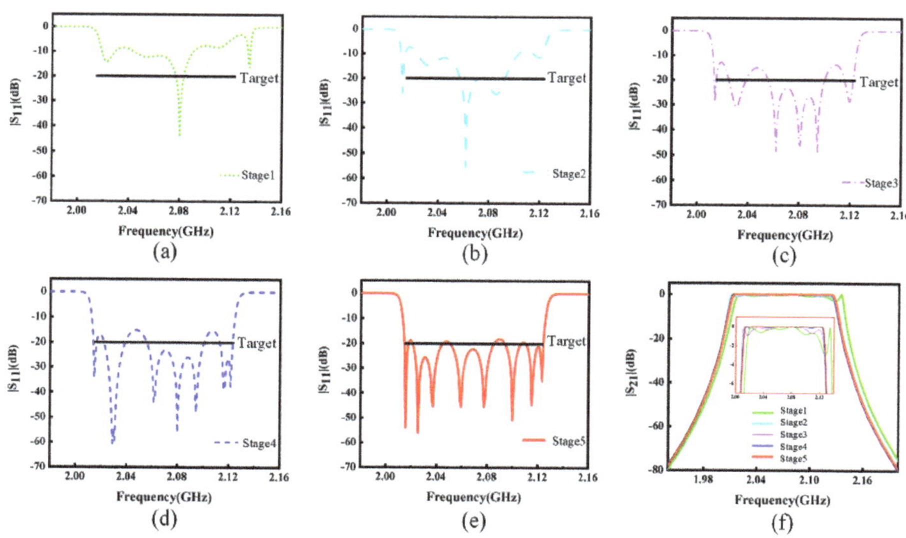

Figure 14. Eighth−order cavity filter HFSS verification of the optimized response in PSO iterations. (**a**–**e**) Reflection coefficient S_{11}. (**f**) Transmission coefficient S_{21}.

Table 3. Eighth-order cavity filter optimization process.

Stage	Initial State (dB)	Data Collection	Time	Satisfaction Data	Reflection Coefficient Distributed Area (dB)
1	−7.3	485	52.13 h	469	[−10, −7.3]
2	−10	1600	4 min	1577 + 469	[−15.4, −10]
3	−15.4	1600	4 min	1563 + 1577 + 469	[−17, −15.4]
4	−17	3200	6 min	142 + 1563 + 1577 + 469	[−18.6, −17]
5	−18.6	–	–	–	–

5. Conclusions

This paper presents a new microwave filter design approach, AOU-1D-CAE-APSO. The AOU-1D-CAE-APSO model is simple in structure and low in complexity. AOU-1D-CAE can adaptively update the content and network parameters of the dataset. Replacing EM simulation software for data collection reduces computational costs and training time. The selection function is added to automatically delete the data that is not in the collection range. At the same time, PSO performs phased optimization and updates the phase optimal solution in time. By inputting the continuously updated stage optimal solution into the PSO, the final optimal solution is obtained through optimization. The final optimal solution refers to the solution that finally meets the design criteria. Avoid falling into the local optimum and ensure the feasibility of the solution.

Author Contributions: Conceptualization, Y.Z., X.W. and Y.W.; methodology, Y.Z., X.W. and Y.W.; software, Y.Z., X.W., Y.W. and N.Y.; validation, Y.Z., X.W. and Y.W.; formal analysis, Y.Z., X.W. and Y.W.; investigation, Y.Z., X.W. and Y.W.; resources, Y.Z., X.W. and Y.W.; data curation, Y.Z., X.W. and Y.W.; writing—original draft preparation, Y.Z., X.W. and Y.W.; writing—review and editing, Y.Z., X.W., Y.W., L.F. and L.Z.; visualization, Y.Z., X.W. and Y.W.; supervision, Y.Z., X.W., Y.W., L.F. and L.Z.; project administration, Y.Z., X.W. and Y.W.; funding acquisition, Y.Z., X.W. and Y.W. All authors have read and agreed to the published version of the manuscript.

Funding: This work was funded by the National Natural Science Foundation of China (NSFC) under Project No.61761032, No.62161032, Nature science foundation of Inner Mongolia under Contract No. 2019MS06006. This work was funded by the Inner Mongolia Foundation 2020MS05059 and Inner Mongolia Department of Transportation NJ-2017-8. This work was also supported by the Shaanxi Key Laboratory of Deep Space Exploration Intelligent Information Technology under Grant No. 2021SYS-04. This work was also supported by the Research and Development of New Energy Vehicle Product Testing Conditions in China—Hohhot Baotou, 2017.

Conflicts of Interest: The authors declare no conflict of interest.

References

1. Cheng, X.; Yao, Y.; Yu, T.; Chen, Z.; Yu, J.; Chen, X. Analysis and Design of a Low-Cost Circularly Polarized Horn Antenna. *IEEE Trans. Antennas Propag.* **2018**, *66*, 7363–7367. [CrossRef]
2. Koziel, S. Low-Cost Data-Driven Surrogate Modeling of Antenna Structures by Constrained Sampling. *IEEE Antennas Wirel. Propag. Lett.* **2017**, *16*, 461–464. [CrossRef]
3. Koziel, S.; Sigurðsson, A.T. Triangulation-Based Constrained Surrogate Modeling of Antennas. *IEEE Trans. Antennas Propag.* **2018**, *66*, 4170–4179. [CrossRef]
4. Feng, F.; Na, W.; Liu, W.; Yan, S.; Zhu, L.; Zhang, Q.-J. Parallel Gradient-Based EM Optimization for Microwave Components Using Adjoint- Sensitivity-Based Neuro-Transfer Function Surrogate. *IEEE Trans. Microw. Theory Tech.* **2020**, *68*, 3606–3620. [CrossRef]
5. Kalantari, L.S.; Bakr, M.H. Wideband Cloaking of Objects with Arbitrary Shapes Exploiting Adjoint Sensitivities. *IEEE Trans. Antennas Propag.* **2016**, *64*, 1963–1968. [CrossRef]
6. Khalatpour, A.; Amineh, R.K.; Cheng, Q.S.; Bakr, M.H.; Nikolova, N.K.; Bandler, J.W. Accelerating Space Mapping Optimization with Adjoint Sensitivities. *IEEE Microw. Wirel. Compon. Lett.* **2011**, *21*, 280–282. [CrossRef]
7. Zhang, C.; Feng, F.; Gongal-Reddy, V.; Zhang, Q.J.; Bandler, J.W. Cognition-Driven Formulation of Space Mapping for Equal-Ripple Optimization of Microwave Filters. *IEEE Trans. Microw. Theory Tech.* **2015**, *63*, 2154–2165. [CrossRef]
8. Gregory, M.D.; Bayraktar, Z.; Werner, D.H. Fast Optimization of Electromagnetic Design Problems Using the Covariance Matrix Adaptation Evolutionary Strategy. *IEEE Trans. Antennas Propag.* **2011**, *59*, 1275–1285. [CrossRef]

9. Koshka, Y.; Novotny, M.A. Comparison of D-Wave Quantum Annealing and Classical Simulated Annealing for Local Minima Determination. *IEEE J. Sel. Areas Inf. Theory* **2020**, *1*, 515–525. [CrossRef]
10. Li, C.; Yang, S.; Nguyen, T.T. A Self-Learning Particle Swarm Optimizer for Global Optimization Problems. *IEEE Trans. Syst. Man Cybern. Part B (Cybern.)* **2012**, *42*, 627–646. [CrossRef]
11. Choi, K.; Jang, D.; Kang, S.; Lee, J.; Chung, T.; Kim, H. Hybrid Algorithm Combing Genetic Algorithm with Evolution Strategy for Antenna Design. *IEEE Trans. Magn.* **2016**, *52*, 7209004. [CrossRef]
12. Januszkiewicz, Ł.; di Barba, P.; Jopek, L.; Hausman, S. Many-objective automated optimization of a four-band antenna for multiband wireless sensor networks. *Sensors* **2018**, *18*, 3309. [CrossRef]
13. Januszkiewicz, Ł.; di Barba, P.; Hausman, S. Multi-objective optimization of a wireless body area network for varying body positions. *Sensors* **2018**, *18*, 3406. [CrossRef]
14. Dong, J.; Qin, W.; Wang, M. Fast Multi-Objective Optimization of Multi-Parameter Antenna Structures Based on Improved BPNN Surrogate Model. *IEEE Access* **2019**, *7*, 77692–77701. [CrossRef]
15. Xia, B.; Pham, M.; Zhang, Y.; Koh, C. A Global Optimization Algorithm for Electromagnetic Devices by Combining Adaptive Taylor Kriging and Particle Swarm Optimization. *IEEE Trans. Magn.* **2013**, *49*, 2061–2064. [CrossRef]
16. Deng, Z.; Rotaru, M.D.; Sykulski, J.K. Kriging Assisted Surrogate Evolutionary Computation to Solve Optimal Power Flow Problems. *IEEE Trans. Power Syst.* **2020**, *35*, 831–839. [CrossRef]
17. Ciccazzo, A.; di Pillo, G.; Latorre, V. A SVM Surrogate Model-Based Method for Parametric Yield Optimization. *IEEE Trans. Comput. -Aided Des. Integr. Circuits Syst.* **2016**, *35*, 1224–1228. [CrossRef]
18. Liu, B.; Yang, H.; Lancaster, M.J. Global Optimization of Microwave Filters Based on a Surrogate Model-Assisted Evolutionary Algorithm. *IEEE Trans. Microw. Theory Tech.* **2017**, *65*, 1976–1985. [CrossRef]
19. Liu, B.; Akinsolu, M.O.; Song, C.; Hua, Q.; Excell, P.; Xu, Q.; Huang, Y.; Imra, M.A. An Efficient Method for Complex Antenna Design Based on a Self Adaptive Surrogate Model-Assisted Optimization Technique. *IEEE Trans. Antennas Propag.* **2021**, *69*, 2302–2315. [CrossRef]
20. Cai, X.; Gao, L.; Li, X. Efficient Generalized Surrogate-Assisted Evolutionary Algorithm for High-Dimensional Expensive Problems. *IEEE Trans. Evol. Comput.* **2020**, *24*, 365–379. [CrossRef]
21. Zhang, Z.; Cheng, Q.S.; Chen, H.; Jiang, F. An Efficient Hybrid Sampling Method for Neural Network-Based Microwave Component Modeling and Optimization. *IEEE Microw. Wirel. Compon. Lett.* **2020**, *30*, 625–628. [CrossRef]
22. Yu, X.; Hu, X.; Liu, Z.; Wang, C.; Wang, W.; Ghannouchi, F.M. A Method to Select Optimal Deep Neural Network Model for Power Amplifiers. *IEEE Microw. Wirel. Compon. Lett.* **2021**, *31*, 145–148. [CrossRef]
23. Feng, F.; Na, W.; Liu, W.; Yan, S.; Zhu, L.; Ma, J.; Zhang, Q.-J. Multifeature-Assisted Neuro-transfer Function Surrogate-Based EM Optimization Exploiting Trust-Region Algorithms for Microwave Filter Design. *IEEE Trans. Microw. Theory Tech.* **2020**, *68*, 531–542. [CrossRef]
24. Feng, F.; Zhang, C.; Ma, J.; Zhang, Q. Parametric Modeling of EM Behavior of Microwave Components Using Combined Neural Networks and Pole-Residue-Based Transfer Functions. *IEEE Trans. Microw. Theory Tech.* **2016**, *64*, 60–77. [CrossRef]
25. Feng, F.; Zhang, C.; Zhang, S.; Gongal-Reddy, V.-M.-R.; Zhang, Q.-J. Parallel EM optimization approach to microwave filter design using feature assisted neuro-transfer functions. In Proceedings of the 2016 IEEE MTT-S International Microwave Symposium (IMS), San Francisco, CA, USA, 22–27 May 2016; pp. 1–3. [CrossRef]
26. Zhang, Y.L.; Wang, Y.X.; Yi, Y.X.; Wang, J.L.; Liu, J.; Chen, Z.X. Coupling Matrix Extraction of Microwave Filters by Using One-Dimensional Convolutional Autoencoders. *Front. Phys.* **2021**, *9*, 716881. [CrossRef]
27. Zhao, P.; Wu, K. Homotopy Optimization of Microwave and Millimeter-Wave Filters Based on Neural Network Model. *IEEE Trans. Microw. Theory Tech.* **2020**, *68*, 1390–1400. [CrossRef]
28. Zhang, Y.L.; Su, T.; Li, Z.P.; Liang, C.H. A hybrid computeraided tuning method for microwave filters. *Prog. Electromagn. Res.* **2013**, *139*, 559–575. [CrossRef]
29. Zhang, Y.L.; Su, T.; Li, Z.P.; Liang, C.H. A fast tuning method for microwave fifilter using VF-ASM technology. *Prog. Electromagn. Res.* **2013**, *30*, 25–37. [CrossRef]

electronics

MDPI

Article

Joint Analysis of Front-Door and Back-Door Couplings of PIN Limiter Based on Improved Equivalent Circuit Model

Tao Liu [1], Le Xu [1,*], Qiwei Li [2], Bin Yao [1] and Xiaowei Shi [1]

[1] School of Electronic Engineering, Xidian University, Xi'an 710071, China
[2] China Academy of Space Technology, Xi'an 710129, China
* Correspondence: lexu@mail.xidian.edu.cn

Abstract: Based on the previous research on electromagnetic pulse coupling, which pays more attention to the front-door coupling or the back-door coupling alone, this paper analyzes the influence of an electromagnetic pulse on electronic devices and systems through the joint analysis of front-door and back-door couplings using the finite-difference time-domain method (FDTD). This specific measure is used to simplify the front-door coupling to the voltage source injection, which occurs simultaneously with plane wave irradiation. This coupling scheme of the front door and back door with the voltage source and plane wave acting simultaneously is rarely seen in previous analyses, which also gives consideration to the working state of the circuit. Although the equivalent circuit model is widely used, it cannot effectively reflect the working state of the diode circuit under the conditions of large injection and high frequency. In view of the limited application scenarios of the traditional equivalent circuit model, which cannot accurately describe the internal response characteristics of the diode under different electromagnetic pulse coupling, this paper introduces an improved equivalent circuit model based on the physical model. Taking the Positive Intrinsic-Negative (PIN) limiter as the target, this paper analyzes the influence of the front-door and back-door joint coupling on its performance under different electromagnetic pulses and then gives protection suggestions.

Keywords: equivalent circuit model; physical model; limiter; diode; electromagnetic coupling; FDTD; field-circuit coupling

Citation: Liu, T.; Xu, L.; Li, Q.; Yao, B.; Shi, X. Joint Analysis of Front-Door and Back-Door Couplings of PIN Limiter Based on Improved Equivalent Circuit Model. *Electronics* **2022**, *11*, 3921. https://doi.org/10.3390/electronics11233921

Academic Editor: Anna Richelli

Received: 26 October 2022
Accepted: 24 November 2022
Published: 28 November 2022

1. Introduction

With the development of high-precision integrated circuits and the rapid development of electronic science and technology, the components of electronic equipment are becoming more intensive [1], the working frequency of the equipment is gradually rising, and the required power consumption is gradually declining. At the same time, many hidden dangers are becoming obvious [2,3], such as the increasingly complex electromagnetic environment and the higher electromagnetic sensitivity of communication equipment to the outside world.

The way in which electromagnetic pulse affects electronic equipment is mainly divided into front-door coupling and back-door coupling [4–6]. Front-door coupling enters the device through an antenna. Back-door coupling refers to the coupling of the electromagnetic pulse into the equipment by cables connected between different devices or slots in the system.

A large number of scholars have conducted research on the coupling effect of electromagnetic pulses on electronic systems [7,8]. Frank Sabath set up a short pulse simulator for susceptibility investigations, which consists of a compact ultrawide-band source and a half-impulse radiating antenna [9]. For the front-door coupling, the antenna is modeled by different full-wave numerical methods. Simulation usually consumes a lot of computing time and memory because the antenna needs a more refined grid to simulate its electromagnetic characteristics. In order to show the influence of different electromagnetic pulses on antenna coupling, a large number of repeated simulations is needed. Compared with the front-door coupling, the back-door coupling receives more concern. For example, there is a lot of theoretical and numerical research work on the shielding effectiveness of metal cavities with different forms of slots [10]. Mats Backstrom studied the field-to-wire coupling problem and regarded the wire as a receiving antenna [11]. As a full-wave electromagnetic algorithm in the time domain, FDTD is widely used to analyze the coupling between the electromagnetic pulse and the system. For the cable-coupling problem, the hybrid FDTD and transmission line method are usually used to analyze the coupling effect of transmission cables in different scenarios. In recent years, in order to analyze the electromagnetic coupling between cables and circuits, the SPICE model has been widely used because it can be used to simulate many circuit devices, such as transistors, capacitors, and inductors [12,13]. However, this method only establishes the SPICE model of the cable or lumped circuit and does not take into account the fact that the electronic system contains an antenna and cable in a real situation. That is, it does not take into account front-door coupling and back-door coupling together. In 2021, some scholars used the hybrid FDTD-SPICE method to analyze the coupling effect of wireless communication systems, including antennas, shielded cables, and metal-shielded cavities with slots under external EMP signals [14]. However, the SPICE model has its own defects: Dan Ren et al. found that the diode had reverse recovery and junction capacitance, which will produce high-order harmonic components at high frequencies, causing conducted interference and radiated interference, leading to electromagnetic compatibility problems. They extracted the diode model according to the measurement results and optimization algorithm, which is used to improve the traditional SPICE model [15]. Furthermore, the SPICE model cannot truly reflect the movement of internal carriers of semiconductor devices under the irradiation of the electromagnetic pulse and the electromagnetic characteristics displayed at the macro level; thus, it is necessary to introduce an improved equivalent circuit model based on a physical model.

Therefore, this paper takes the limiter as the target, introduces the improved equivalent circuit model based on the physical model, analyzes the influence of front-door and back-door joint coupling on its coupling power and limiting performance, and then puts forward protection suggestions for front-door and back-door coupling, respectively. Common measures to protect equipment from electromagnetic interference include electromagnetic shielding, grounding, and filtering. The specific actions include using non-metallic fiber-optic cables when possible and employing methods to decrease the resonance characteristics of critical equipment enclosures [16]. Frank Leferink has conducted a lot of work on electromagnetic interference and proposed lots of valuable advice, such as limiting the length of exposed cable and placing exposed cables out of the line of sight of transmitters above 400 MHz when common mode currents on cables are exposed to high-intensity radiated fields and a nuclear electromagnetic pulse [17].

The structure and main contribution of this paper are as follows. Section 2 introduces the SPICE model of the limiter and the improved equivalent circuit model based on the physical model. In Section 3, based on the improved equivalent circuit model, the joint-coupling characteristics of the front door and back door of the limiter are analyzed, and the protection suggestions of front-door limiting and back-door shielding are given. Section 4 is a summary. A flow chart illustrates the structure and main contributions of this paper in Figure 1.

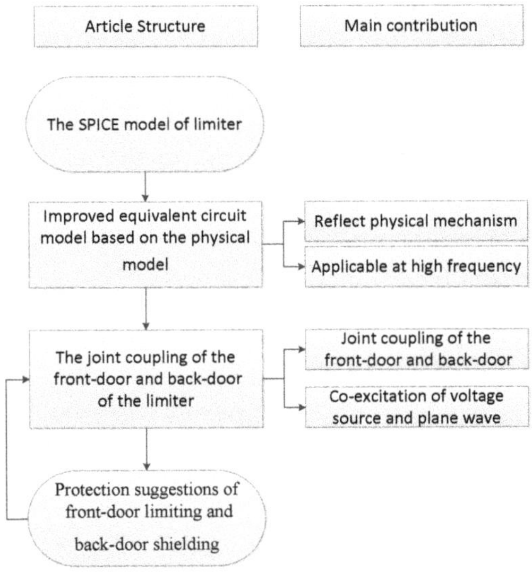

Figure 1. Flow chart for the structure and main contributions of this paper.

2. Theory and Model

This section first introduces the SPICE model of the limiter and gives the traditional equivalent circuit formula of the diode. Then, the diode simulation method based on the physical model, as well as the resulting improved equivalent circuit model, is described. Finally, we give the joint coupling scheme of the front-door and back-door of the limiter.

2.1. Traditional SPICE Model of Limiter

The common SPICE model of the single-stage and single-tube limiter is shown in Figure 2. R_A and R_L are equivalent line resistance and load impedance, both set to 50 Ω.

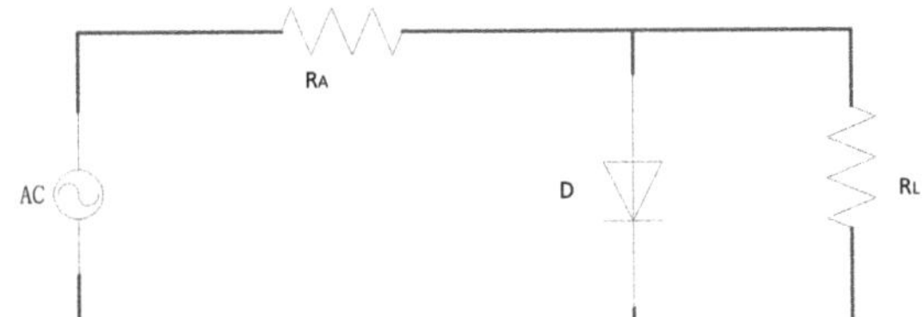

Figure 2. Single-stage and single-tube limiter.

As shown in Figure 3, the diode is located between nodes (i, j, k) and (i, j, k + 1). The current direction is the z direction, and I_d can be described by the following formula

$$I = I_d[e^{qV_d/kT} - 1] \quad (1)$$

where I_d is the dark current, q is the absolute value of electron charge, k is the Boltzmann constant, V_d is the voltage across the diode, and T is the thermodynamic temperature.

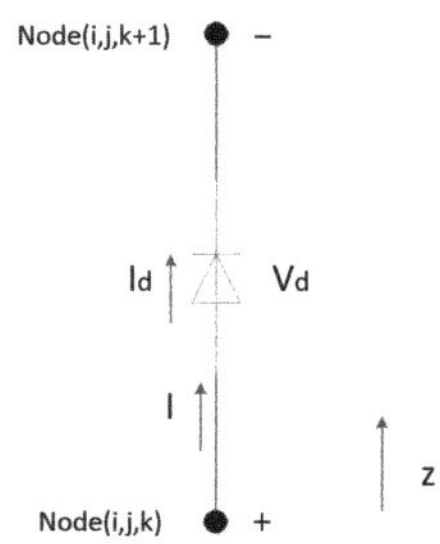

Figure 3. Schematic diagram of the diode.

2.2. Diode Formula Based on Physical Model

The traditional equivalent circuit formula cannot accurately reflect the device characteristics, especially in the case of high frequency, so the physical model is introduced. According to the theory of semiconductor devices, the basic equation consists of three parts: the carrier transport equation, continuity equation, and Poisson equation, which are summarized into a set of basic differential equations for the analysis of semiconductor devices:

$$\frac{\partial p}{\partial t} = -\frac{1}{q}\nabla \cdot \mathbf{J_p} + G_p - U_p \tag{2}$$

$$\frac{\partial n}{\partial t} = -\frac{1}{q}\nabla \cdot \mathbf{J}_n + G_n - U_n \tag{3}$$

$$\mathbf{J_p} = -qD_p\nabla p - q\mu_p p\nabla\psi \tag{4}$$

$$\mathbf{J_n} = -qD_n\nabla n - q\mu_n n\nabla\psi \tag{5}$$

$$\nabla\cdot\nabla\psi = -\frac{q}{\varepsilon}(N_d - N_a + p - n) \tag{6}$$

where ε, $\mathbf{J_n}$, ψ, $\mathbf{J_p}$, n, p, N_a, q, N_d, and t are the permittivity, electron current density, electrostatic potential, hole current density, electron concentrations, hole concentrations, hole doping concentrations, electronic charge, electron doping concentrations, and time, respectively. G_n and G_p represent carrier generation rates for the electron and hole. U_n and U_p are recombination rates for the electron and hole. D_p and D_n are the hole–diffusion coefficient and electron–diffusion coefficient.

According to the diode response curve fitted by the physical model, the improved equivalent circuit formula can be proposed [18]:

$$\begin{aligned} I = \varepsilon(f_0 - f)\cdot I_0[e^{\frac{q}{kT}V_d} - 1] + \varepsilon(f - f_0)\Big\{\alpha\cdot\tfrac{1}{2\pi f L_0}V_d \\ +\beta\cdot[\varepsilon(V_d - V_1)2\pi f C_1\cdot V_d + \varepsilon(V_1 - V_d)2\pi f C_2\cdot V_d]\Big\} \end{aligned} \tag{7}$$

where

$$\alpha = \frac{1}{1 + e^{a(V_d - V_0)}} \tag{8}$$

$$\beta = \frac{1}{1 + e^{-bV_d}} \tag{9}$$

where α and β are control coefficients, I represents the current flowing through the diode, I_0 represents the saturation current, q is the absolute value of the electron charge, k is the Boltzmann constant, and T is the thermodynamic temperature. V_d is the voltage across the diode, and V_0 is the empirical voltage value. f_0 is the empirical value; f represents the frequency of solving the problem; $\varepsilon(f_0 - f)$, $\varepsilon(f - f_0)$, $\varepsilon(V_d - V_1)$ and $\varepsilon(V_1 - V_d)$ represent the step function; L_0 is the empirical inductance value; C_1 and C_2 are empirical values; and V_1 represents the startup voltage of the diode.

The physical meaning of the formula is defined as follows: when the frequency f is lower than f_0, it is calculated according to the traditional empirical formula. When f is higher than f_0, the traditional empirical formula is no longer applicable. At high frequency and low voltage, the diode presents high impedance characteristics, and at high frequency and high voltage, the diode presents low impedance characteristics, and these are respectively used to control the diode response of low power input (equivalent to low input voltage) and high power input (equivalent to high input voltage) at high frequency.

2.3. Joint Analysis of Front-Door and Back-Door Couplings of PIN Limiter Based on Physical Model

In Sections 2.1 and 2.2, we introduce the limiter model based on the traditional equivalent circuit formula and the improved equivalent circuit formula based on the physical model. On this basis, we introduce how to create the front-door and back-door joint coupling of the limiter based on the physical model.

The system used in this paper consists of a receiving antenna and a limiter, which are placed in an equipment cabin with a slot, as shown in Figure 4. In order to study the influence of front-door and back-door coupling on the working characteristics of the limiter circuit, the following simplification is made: As introduced in the literature [14], there is an analytical relationship between the signal received by the antenna from the front door and the electromagnetic pulse, so the signal reaching the limiter can be expressed as a certain form of excitation source. This paper simplifies the front-door coupling through the antenna as the sine wave to the voltage source. The back-door coupling usually includes electromagnetic wave components from various frequencies in space, so the form of the plane wave adopts a Gaussian pulse. In order to maximize the effect of back-door coupling, a bare PIN circuit without any protective measures is adopted here. Here, we use both voltage source and plane wave excitation so as to simulate the joint coupling characteristics of the front door and back door under different electromagnetic pulses. Then, we give the protection scheme of front-door limiting and back-door shielding accordingly.

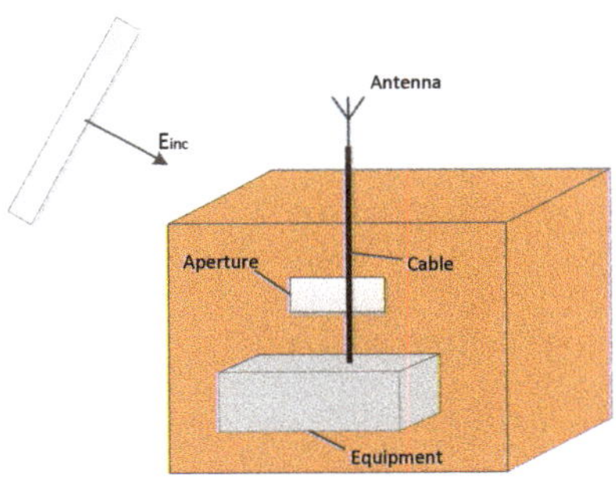

Figure 4. Typical wireless communication system illuminated by an electromagnetic pulse.

3. Numerical Simulation and Analysis

Based on the improved equivalent circuit formula extracted from the physical model proposed in the previous section, this section analyzes the joint coupling characteristics of the front door and back door of the limiter under the irradiation of the electromagnetic pulse and gives protection suggestions accordingly.

3.1. Simulation Effectiveness Verification

In order to verify the effectiveness of the simulation method, this section verifies the electromagnetic characteristics of the target under plane wave irradiation and voltage injection.

3.1.1. Front-Door Injection

The accuracy of the circuit response curve under voltage injection is verified below.

A. Traditional SPICE simulation

The response curve of the diode is obtained by the FDTD method, as shown in Figure 5, using the same simulation condition as reference [18]. When the source voltage is higher than 0.7 V, the sampling voltage of the diode remains at 0.7 V, which is consistent with the theoretical analysis [19].

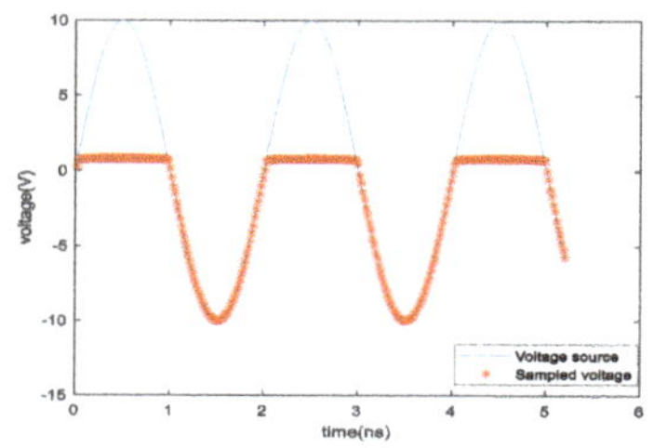

Figure 5. Diode response curve.

B. Improved equivalent circuit

Now, we start the simulation of the circuit shown in Figure 2 with the improved equivalent circuit formula. The voltage at both sides of the diode is observed. Here, we define f_0 = 2 GHz, V_0 = 10 V, L_0 = 6 H, a = 2, b = 2, C_1 = 100 pF, C_2 = 1 pF, and V_1 = 0.7 V. Three examples are given to verify the effectiveness of the improved equivalent circuit method.

- f = 400 MHz, V_m = 4 V

In the first example, we define the frequency of solving the problem as 400 MHz and the maximum amplitude of the excitation source as 4 V. It is used for testing the improved equivalent circuit formula under low frequency. The results are shown in Figure 6, which are consistent with the analysis in reference [18].

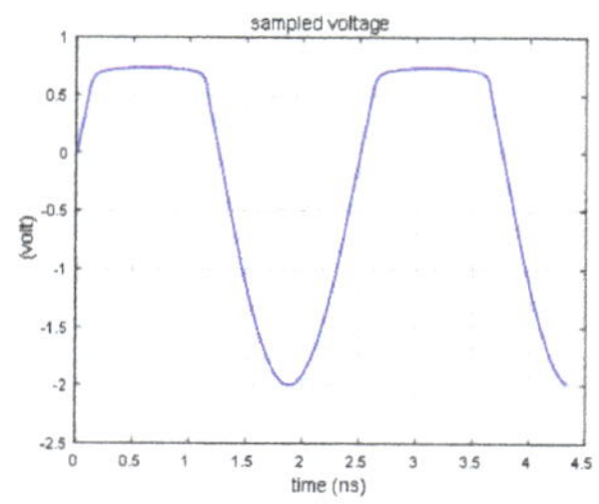

Figure 6. Result of improved equivalent circuit method under low frequency.

As we can see from the picture, it is consistent with the traditional equivalent circuit method.

- f = 4 GHz, V_m = 6 V

In the second example, we define the frequency of solving the problem as 4 GHz and the maximum amplitude of the excitation source as 6 V. It is used for testing the improved equivalent circuit formula under high-frequency and low-voltage injection. The results are shown in Figure 7, which are consistent with the analysis in reference [18].

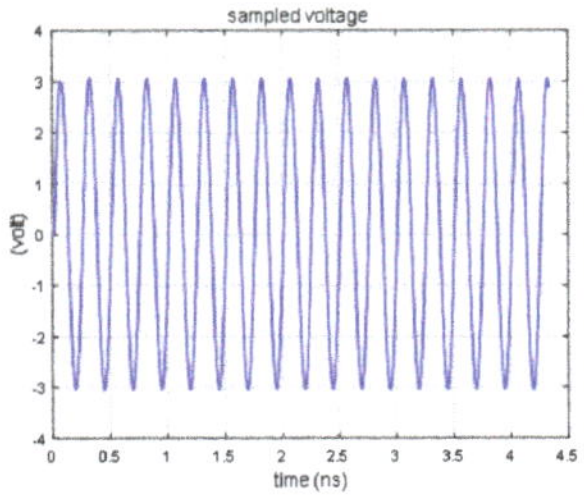

Figure 7. Result of improved equivalent circuit method under high-frequency and low-voltage injection.

The improved equivalent circuit method can feed back the characteristics similar to the physical model at high-frequency and low-voltage injection.

- f = 4 GHz, V_m = 30 V

In the third example, we define the frequency of solving the problem as 4 GHz and the maximum amplitude of the excitation source as 30 V. It is used for testing the improved equivalent circuit formula under high-frequency and high-voltage injection. The results are shown in Figure 8, which are consistent with the analysis in reference [18].

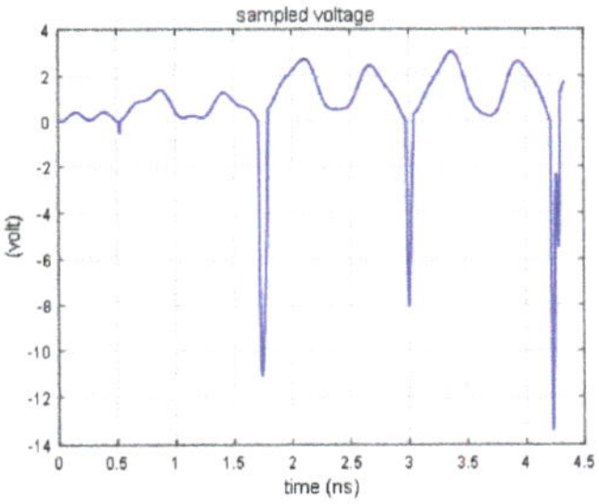

Figure 8. Result of improved equivalent circuit method under high-frequency and high-voltage injection.

As we can see from Figure 8, the improved equivalent circuit method can work at high frequency and high power.

3.1.2. Back-Door Irradiation

For testing the accuracy of the electromagnetic characteristics of the target under the plane wave irradiation, the plane wave is used to irradiate the dielectric sphere, the waveform of which is the Gaussian pulse. The propagation direction is the z direction, and the electric field direction is the x direction. The radius of the dielectric sphere is 10 cm, and the electromagnetic parameters are epsilon_r = 3, sigma_ e = 0.5 S/m, sigma_ m = 0.5 H/m, and mu_r = 2. The spatial step size is dx = dy = dz = 1 cm. We select the center position of the medium sphere as the reference point and obtain the comparison result with CST, as shown in Figure 9.

It can be seen that the FDTD simulation results compare well with CST, which proves the effectiveness of the FDTD algorithm in simulating the near-field coupling problem under plane wave irradiation.

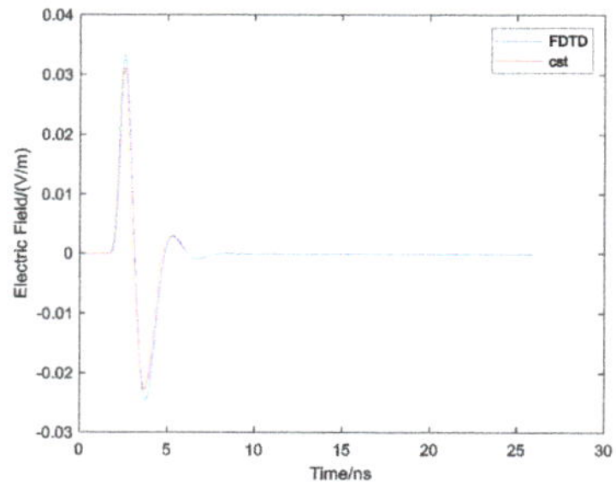

Figure 9. Comparison of field between FDTD and CST.

3.2. Joint Coupling Characteristics of Front Door and Back Door under Different Electromagnetic Pulse Irradiation

Based on Section 2.3, we start the process of joint analysis of front-door and back-door coupling. Firstly, no protective measures are taken for the front door and back door. The circuit used is shown in Figure 2. Because the limiter pays more attention to the output power, the final obvious target is the power on the load impedance. Under the same conditions, the voltage at both ends of the diode can be observed equivalently.

- Low-frequency condition

The voltage of the front-door coupling is mainly a low-frequency component, assuming that the frequency of the sine wave voltage is 400 MHz and the amplitude is 4 V. The amplitude of the Gaussian pulse of the back-door coupling is 1000 V/m, the propagation direction is along the z direction, and the electric field direction is the x direction. The mesh step is dx = dy = dz = 2 mm, and the model is shown in Figure 10. Setting the circuit size to 20 mm × 20 mm × 10 mm, we can observe the voltage at both ends of the diode and the field at the center.

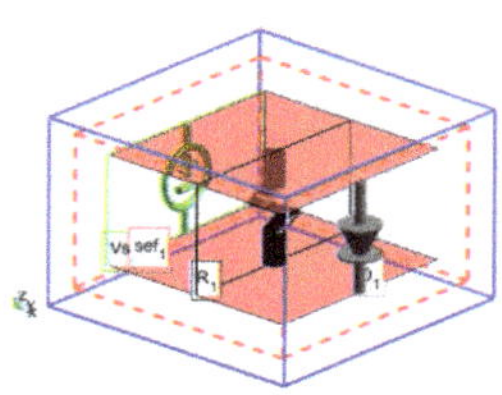

Figure 10. Model of single tube limiter.

The voltage at both ends of the diode is obtained by simulation.

In Figure 11, the influence of the applied Gaussian pulse on the voltage at both ends of the diode can be clearly seen, which makes the voltage–response curve different from that of front-door coupling only in Figure 6. We can see that a pulse whose absolute amplitude takes over 3 V has been introduced. In fact, this is a typical field-circuit coupling problem. As the diode in Figure 3 is located between the node (i, j, k) and the node (i, j, k + 1), the port voltage V_d can be deduced from the electric field path integral when there is no circuit excitation source:

$$V_d = \int_{k}^{k+1} \vec{E_d} \cdot dl \tag{10}$$

where E_d represents the coupling voltage caused by back-door coupling at both ends of the diode [20]. When the front-door and back-door coupling occurs at the same time, the voltage at both ends of the diode is the composition of the coupling voltage caused by back-door irradiation and the voltage caused by the excitation source injection of the front-door circuit. This explains the difference between the coupling voltage waveform at both ends of the diode when the front door and rear door are coupled together and the coupling-voltage waveform when only the front door is injected.

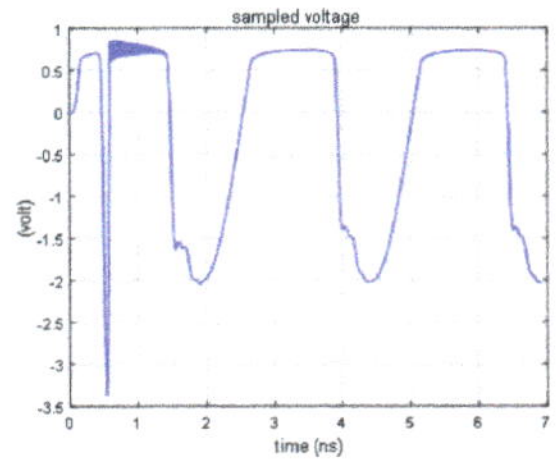

Figure 11. Voltage at both ends of diode at low frequency.

The dynamic changes of the electric field in the circuit are shown in Figure 12 below. These figures are all made by MATLAB 2018 using the animation function.

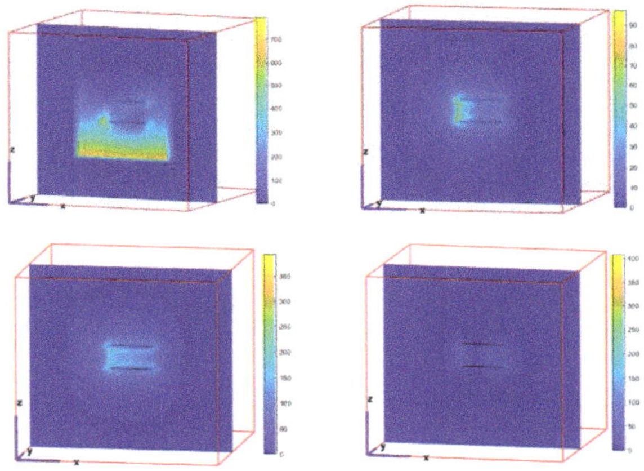

Figure 12. The dynamic changes of the electric field at low frequency.

- High-frequency condition

When the voltage of the front-door coupling is mainly a high-frequency component, it is assumed that the frequency of the sine wave is 4 GHz, the amplitude is 6 V, and the other settings are the same as those in low frequency before. The updated formula of the diode is in accordance with the improved equivalent circuit formula, and the voltage at both ends of the diode is simulated.

In Figure 13, the influence of the applied Gaussian pulse on the voltage at both ends of the diode can be clearly seen, which makes the voltage–response curve different from that of front-door coupling only in Figure 7. We can see that a pulse whose absolute amplitude takes over 5 V has been introduced. This can be explained as above. The dynamic changes of the electric field in the circuit are shown in the Figure 14 below.

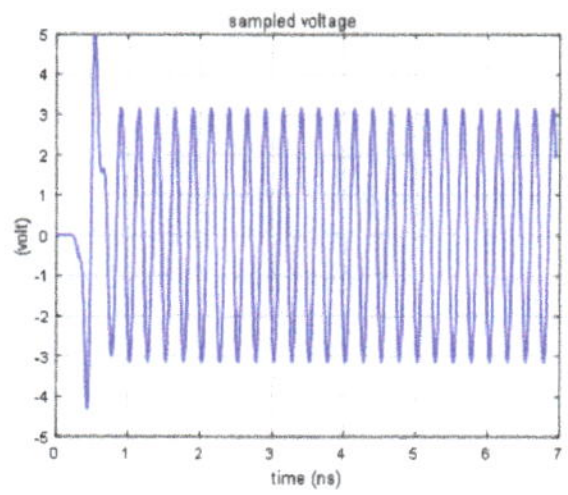

Figure 13. Voltage at both ends of diode at high frequency.

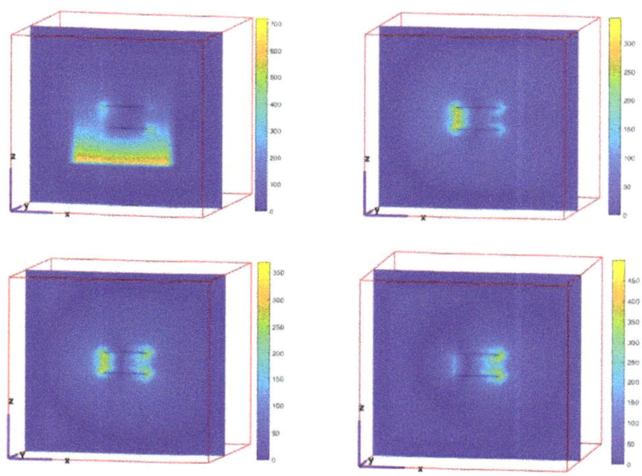

Figure 14. The dynamic changes of the electric field at high frequency.

3.3. Joint Coupling Characteristics of Front-Door and Back-Door Coupling after Taking Protection Measures

In order to reduce the electromagnetic coupling of the front door and back door, this paper adopts the protection scheme of front-door limiting and back-door shielding.

3.3.1. Single-Stage and Double-Tube Limiter

In order to achieve a better limiting effect of the front door, the single-tube limiter is changed to the double-tube, as shown in Figure 15.

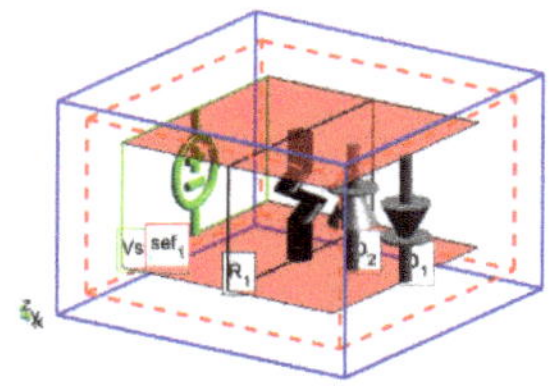

Figure 15. Double-tube limiter.

1. Low-frequency condition

The simulation setting is the same as when the frequency of the front-door coupling is 400 MHz, and the voltage at both ends of the diode is shown in Figure 16.

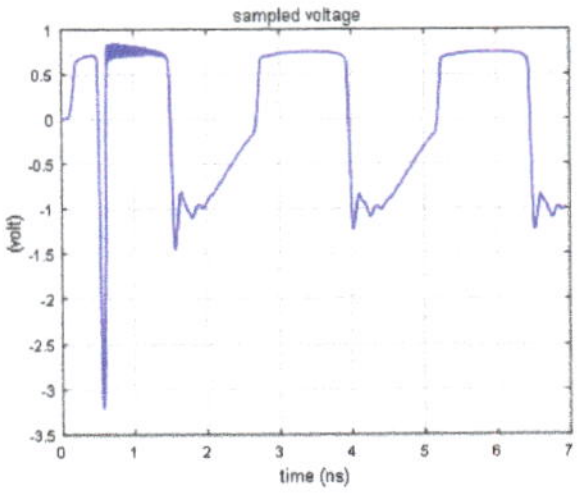

Figure 16. Voltage at both ends of the double-tube under low frequency.

Compared with Figure 11, it is found that the coupling voltage of the front door decreases significantly, which means that the output power decreases significantly. However, due to the lack of back-door protection, the back-door coupling voltage is still obvious. The dynamic changes of the electric field in the circuit are shown in the Figure 17 below.

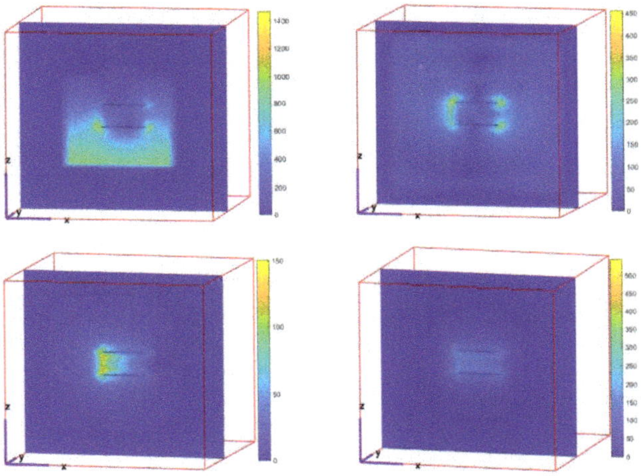

Figure 17. The dynamic changes of the electric field with double-tube under low frequency.

2. High-frequency condition

Setting the front-door coupling frequency to 4 GHz, the other settings are consistent with those at low frequency. The voltage at both ends of the diode of the double-tube limiter can be obtained in Figure 18.

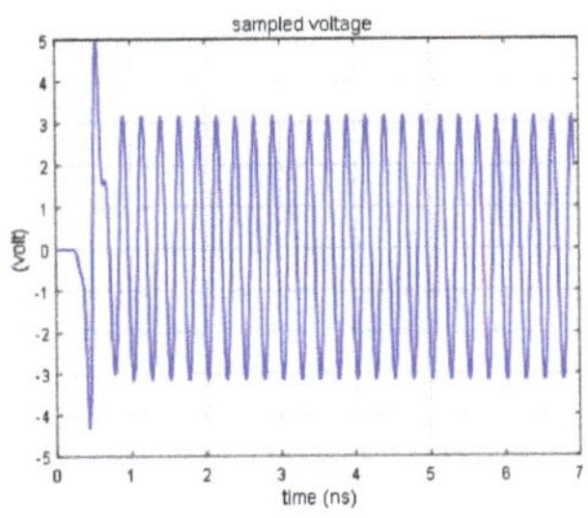

Figure 18. Voltage at both ends of double-tube under high frequency.

Compared with Figure 13, when the front door is injected with a high frequency and low voltage, the diode shows high impedance, which is equivalent to a large inductance. After the two tubes are connected in parallel, it is still a large inductance, so the coupling waveform changes little compared with Figure 13. The dynamic changes of the electric field in the circuit are shown in the Figure 19 below.

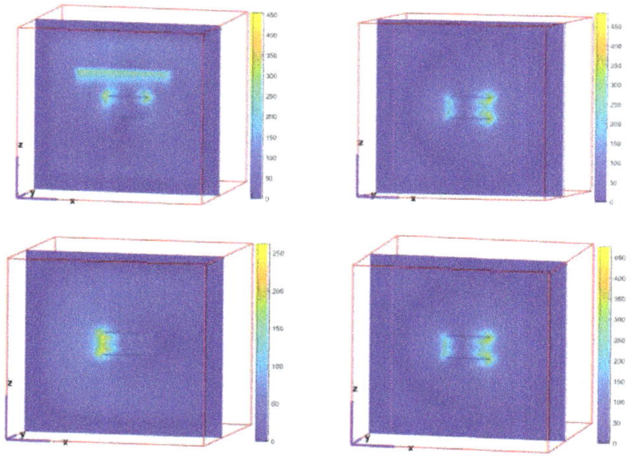

Figure 19. The dynamic changes of the electric field with double-tube under high frequency..

3.3.2. Limiter with Shielding Cavity

In order to reduce the influence of the back-door coupling on the performance of the limiter, the limiter is placed in a PEC metal shielding cavity. The size of the shielding cavity is 40 mm × 40 mm × 20 mm, and the wall thickness is 4 mm. In reality, it is difficult to achieve complete shielding, so a 4 mm × 4 mm × 4 mm hole is opened at the bottom of the shielding cavity to simulate this situation. The model is shown in Figure 20.

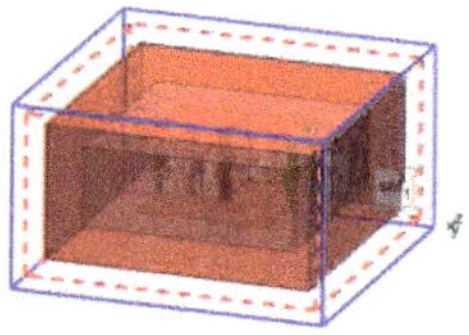

Figure 20. Shielding cavity with built-in limiter.

3. Low-frequency condition

The simulation setting is the same as that of the double-tube limiter at low frequency, and the voltage at both ends of the diode is obtained in Figure 21.

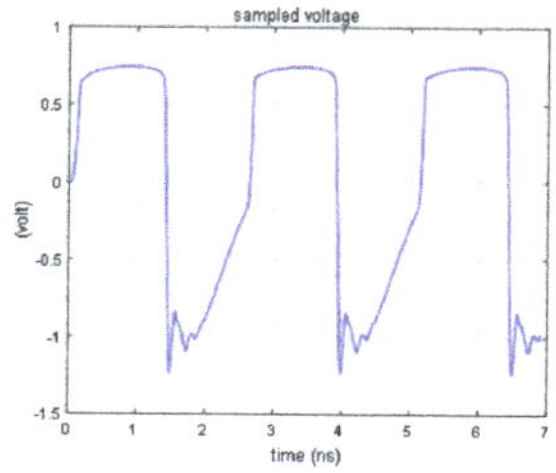

Figure 21. Voltage at both ends of diode with shielding cavity under low frequency.

Compared with Figure 16, it can be seen that due to the skin effect of the shielding cavity and the reflection of electromagnetic waves, the back-door coupling component has been suppressed and eliminated, which proves the effectiveness of the shielding cavity. The dynamic changes of the electric field in the circuit are shown in the Figure 22 below.

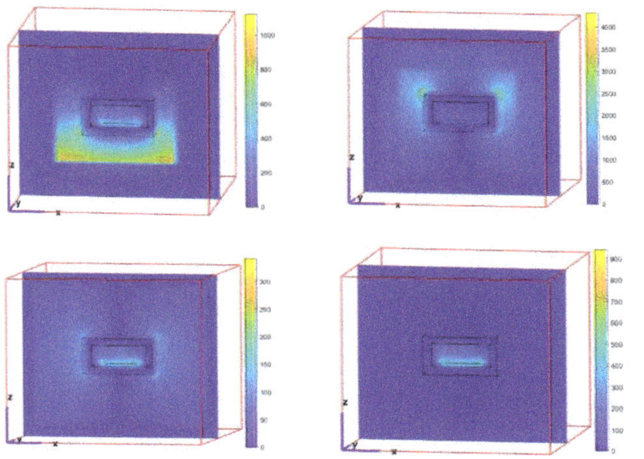

Figure 22. The dynamic changes of the electric field with shielding cavity under low frequency.

From the field calculation results, it is also clear that the back-door coupling is effectively suppressed due to the existence of the shielded cavity, which is consistent with the circuit analysis results.

4. High-frequency condition

The simulation setting is the same as that of the double-tube limiter at high frequency, and the voltage at both ends of the diode is obtained in Figure 23.

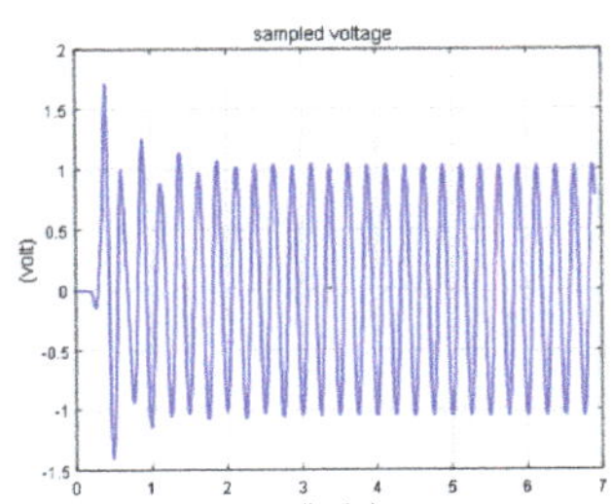

Figure 23. Voltage at both ends of the diode.

Compared with Figure 18, it can be seen that due to the skin effect of the shielding cavity and the reflection of electromagnetic waves, the back-door coupling component has been suppressed and eliminated, which proves the effectiveness of the shielding cavity. The dynamic changes of the electric field in the circuit are shown in the Figure 24 below.

From the field calculation results, it is also clear that the back-door coupling is effectively suppressed due to the existence of the shielded cavity, which is consistent with the circuit analysis results.

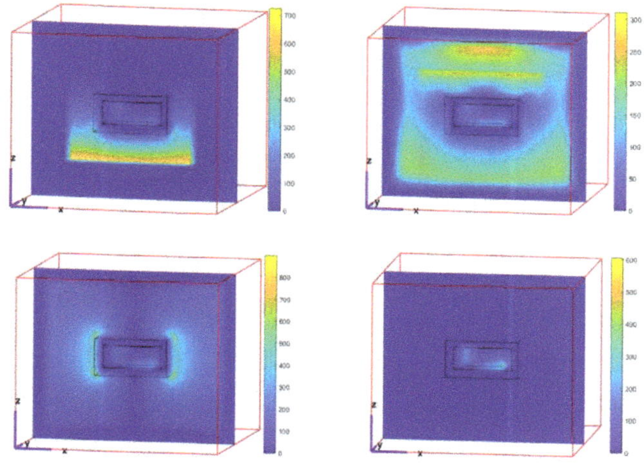

Figure 24. The dynamic changes of the electric field.

4. Conclusions

This paper studies the joint coupling effect of the front door and back door of the PIN limiter under electromagnetic pulse irradiation based on the physical model. Previous research on electromagnetic pulse coupling in electronic devices and systems focuses more on front-door or back-door coupling alone; thus, it cannot accurately reflect the situation when electronic devices and systems suffer from front-door and back-door coupling at the same time. More importantly, when front-door and back-door coupling occurs, the internal circuit is in a working state, and there is excitation. This factor should also be considered. In this paper, the front-door coupling is equivalent to the voltage injection, and the back-door coupling adopts the Gaussian pulse as the plane wave source in order to reflect the coupling of various frequency components to the system. This joint-coupling scheme of the front-door and back-door, with the voltage source and plane wave acting simultaneously, is rarely seen in previous analyses. In view of the limited application scenarios of the traditional equivalent circuit model, an improved diode equation based on the physical model is introduced. The improved equivalent circuit method can be used to analyze the working state of the diode circuit under large-injection and high-frequency conditions so as to expand its application scenarios. By adopting the scheme of "front-door limiting" and "back-door shielding", it is found that the voltage, finally coupled to both ends of the limiter diode, has been significantly reduced compared with that before the protection measures were taken. The equivalent understanding is that the power at both ends of the load has been significantly reduced. Therefore, it has obvious practical significance for protecting load devices.

Author Contributions: Conceptualization, B.Y.; methodology, L.X.; software, T.L.; validation, T.L.; formal analysis, T.L.; investigation, Q.L.; resources, L.X.; data curation, L.X.; writing—original draft preparation, T.L.; writing—review and editing, L.X.; visualization, T.L.; supervision, X.S.; project administration, X.S.; funding acquisition, Q.L. All authors have read and agreed to the published version of the manuscript.

Funding: This research was funded by the National Key Laboratory of Science and Technology on Space Microwave, grant number 6142411332211.

Data Availability Statement: Not applicable.

Conflicts of Interest: The authors declare no conflict of interest.

References

1. Liu, X.; Liu, Y.; Weng, X.; Lin, X.; Yang, Y. Fault Injection of Strong Magnetic Pulse in Digital Integrated Circuit. In Proceedings of the 2021 International Conference on Communications, Information System and Computer Engineering (CISCE 2021), Beijing, China, 14–16 May 2021; pp. 164–167.
2. Gutierrez, H.M.; Christoffersen, C.E.; Steer, M.B. An integrated environment for the simulation of electrical, thermal and electromagnetic interactions in high-performance integrated circuits. In Proceedings of the IEEE 8th Topical Meeting on Electrical Performance of Electronic Packaging (Cat. No.99TH8412), San Diego, CA, USA, 25–27 October 1999; pp. 217–220.
3. Huynh, H.A.; Jo, J.-M.; Nah, W.; Kim, S. EMC Qualification Methodology for Semicustom Digital Integrated Circuit Design. *IEEE Trans. Electromagn. Compat.* **2016**, *58*, 1629–1641. [CrossRef]
4. Becerra, J.; Lopez, Z.; Rangel, A.; Vega, F. A Comparison of Fitting Methods for Modeling the Front Door Coupling of Two Nearby Parabolic Antennas. In Proceedings of the 2018 USNC-URSI Radio Science Meeting (Joint with AP-S Symposium), San Diego, CA, USA, 8–13 July 2018; pp. 27–28.
5. Zhang, D.; Zhou, X.; Cheng, E.; Wan, H.; Chen, Y. Investigation on Effects of HPM Pulse on UAV's Datalink. *IEEE Trans. Electro. Compat.* **2020**, *62*, 829–839. [CrossRef]
6. Guo, L.; Xiao, L.; Chen, J.; Yang, M.; Yang, J. Electromagnetic Pulse Coupling Effect Analysis for Outboard Engine System of Vehicle. In Proceedings of the 2020 IEEE MTT-S International Microwave Workshop Series on Advanced Materials and Processes for RF and THz Applications (IMWS-AMP), Chongqing, China, 29–31 November 2021; pp. 1–3.
7. Omri, D.; Aguili, T. Time-domain techniques for electromagnetic coupling analysis of transient excitations of rectangular cavity through slot. *J. Electromagn. Waves Appl.* **2015**, *29*, 1297–1316. [CrossRef]
8. Hu, X.; Qiu, Y.; Xu, Q.; Tian, A.J. Transient Response of Microstrip Patch Antenna Loaded on a Vehicle Platform Illuminated by Electromagnetic Pulse. *Prog. Electromagn. Res. C* **2020**, *104*, 69–84. [CrossRef]
9. Sabath, F.; Nitsch, D.; Jung, M.; Weise, T.H. Weise, Design and setup of a short pulse simulator for susceptibility investigations. *IEEE Trans. Plasma Sci.* **2002**, *30*, 1722–1727. [CrossRef]
10. Konefal, T.; Dawson, J.; Marvin, A.; Robinson, M.; Porter, S. A Fast Circuit Model Description of the Shielding Effectiveness of a Box With Imperfect Gaskets or Apertures Covered by Thin Resistive Sheet Coatings. *IEEE Trans. Electromagn. Compat.* **2006**, *48*, 134–144. [CrossRef]
11. Silfverskiold, S.; Backstrom, M.; Loren, J. Microwave field-to-wire coupling measurements in anechoic and reverberation chambers. *IEEE Trans. Electromagn. Compat.* **2002**, *44*, 222–232. [CrossRef]
12. Fujita, K. MNL-FDTD/SPICE Method for Fast Analysis of Short-Gap ESD in Complex Systems. *IEEE Trans. Electromagn. Compat.* **2016**, *58*, 709–720. [CrossRef]
13. Wang, X.; Wang, L.; Zhuo, J.; Lu, X.; Yuan, M.; Zhou, J.; Liu, Q.H. A Hybrid CN-FDTD-SPICE Solver for Field-Circuit Analyses in Low-Frequency Wideband Problems. *IEEE Trans. Compon. Packag. Manuf. Technol.* **2020**, *10*, 1721–1728. [CrossRef]
14. Hu, X.; Qiu, Y.; Xu, Q.L.; Tian, J. A Hybrid FDTD–SPICE Method for Predicting the Coupling Response of Wireless Communication System. *IEEE Trans. Electromagn. Compat.* **2021**, *63*, 1530–1541. [CrossRef]
15. Ren, D.; Xu, G.; Li, J.-Q.; Pan, Z.-Y.; Zhao, X.; Du, P.-A. Improvement of the SPICE Model of Diode Based on Measurement and Nonlinear Fitting Random Optimization Algorithm. *Electronics* **2022**, *11*, 3461. [CrossRef]
16. IEEE. *IEEE Recommended Practice for Protecting Publicly Accessible Computer Systems from Intentional Electromagnetic Interference (IEMI), IEEE Std 1642-2015*; IEEE: Manhattan, NY, USA, 2015; pp. 1–39.
17. Leersum, B.; Ven, J.-K.; Bergsma, H.; Buesink, F.; Leferink, F. Protection Against Common Mode Currents on Cables Exposed to HIRF or NEMP. *IEEE Trans. Electro. Compat.* **2016**, *58*, 1297–1305. [CrossRef]
18. Liu, T.; Xu, L.; He, Y.; Wu, H.; Yang, Y.; Wu, N.; Yang, X.; Shi, X.; Wei, F. A Novel Simulation Method for Analyzing Diode Electrical Characteristics Based on Neural Networks. *Electronics* **2021**, *10*, 2337. [CrossRef]
19. Elsherbeni, A.; Demir, V. *The Finite Difference Time Domain Method for Electromagnetics: With MATLAB Simulations*; SciTech Publishing: New York, NY, USA, 2016.
20. Zeng, H.; Tang, Y.; Duan, X.; Chen, X. A Physical Model-Based FDTD Field-Circuit Co-Simulation Method for Schottky Diode Rectifiers. *IEEE Access* **2019**, *7*, 87265–87272. [CrossRef]

electronics

MDPI

Article

Research on Testing Method for Shielding Effectiveness of Irregular Cavity Based on Field Distribution Characteristics

Jinjing Ren [1], Yuhao Pan [1], Zhongyuan Zhou [1,*] and Tao Zhang [2,3]

1 Research Center for Electromagnetic Environmental Effects, Southeast University, Nanjing 211189, China
2 Aviation Key Laboratory of Science and Technology on Electromagnetic Environmental Effects, Shenyang 110135, China
3 School of Information Science and Engineering, Southeast University, Nanjing 211189, China
* Correspondence: zyzhou@seu.edu.cn

Abstract: A method of measuring the shielding effectiveness of the irregular cavity is proposed in this paper. An electromagnetic topology model of the irregular cavity is established according to the characteristics of the irregular cavity, and the electromagnetic field distribution characteristics inside the irregular cavity are obtained based on the simulation of the field distribution of the irregular cavity. Combined with the test method of regular cavity shielding effectiveness, the characterization and testing methods of the shielding effectiveness of the irregular cavity are given comprehensively, which compared with the conventional testing method, verifies how effectively the shielding effectiveness of the irregular cavity is tested.

Keywords: irregular cavity; electromagnetic topology; field distribution; shielding effectiveness; test method

Citation: Ren, J.; Pan, Y.; Zhou, Z.; Zhang, T. Research on Testing Method for Shielding Effectiveness of Irregular Cavity Based on Field Distribution Characteristics. *Electronics* **2023**, *12*, 1035. https://doi.org/10.3390/electronics12041035

Academic Editor: Andrea Randazzo

Received: 26 November 2022
Revised: 6 February 2023
Accepted: 16 February 2023
Published: 19 February 2023

1. Introduction

With the electromagnetic environment growing increasingly complex, the electromagnetic environmental effects are receiving more and more attention, and electronic equipment is susceptible to electromagnetic interference or even failure [1]. In order to ensure the reliable operation of the equipment, protective measures against electromagnetic interference are indispensable, and electromagnetic shielding (SE) is one of the most important methods to suppress electromagnetic interference. Usually, a metal cavity is used outside for electromagnetic shielding and protection. Considering the special installation or fixing requirements of some equipment and the arrangement requirements of the shielding cavity, the cavity structure is not a regular cuboid or cylinder, and there may be irregular structures such as protrusions, plates, steps, and blocks. In actual applications, there will be some apertures in the metal cavity due to the requirements of ventilation, heat dissipation, and the wiring of the internal equipment. These apertures will reduce the SE of the cavity because of electromagnetic field coupling. Therefore, research on the shielding effectiveness of irregular cavities with apertures is of great significance to the electromagnetic protection of electronic equipment.

For the research on the testing method of the shielding effectiveness of the irregular cavity, the classification of the irregular cavity is defined firstly. Secondly, an electromagnetic topology model to analyze the shielding effectiveness of the irregular cavity is built, and then the characteristics of the electromagnetic field in the cavity are obtained, and the characterization method of the shielding effectiveness of the irregular cavity is given. After these steps, the testing method for the shielding effectiveness of the irregular cavity is proposed.

There are two main methods for calculating the shielding effectiveness of the cavity with apertures: numerical and analytical ones. Complex structures can be simulated through numerical methods [2–6], but a lot of computing time and memory are usually needed to build a model with sufficient accuracy. At the same time, they are not conducive to the regularity analysis of the effects of parameters. By contrast, the physical meaning obtained using the analytical algorithm [7–10], with a fast calculation speed, is more explicit than the one established by the actual model. At present, there are some analytical methods for the regular of shielding cavities with apertures [11–16], but further simplified ones are still needed, and the calculation of the shielding effectiveness of actual irregular cavities should be summarized and studied. In this paper, the electromagnetic topology model in the analytical method is used to establish the calculation model of the shielding effectiveness of the irregular cavity, and the irregular structure is classified.

For the field distribution research of the shielded cavity, most previous studies chose the center position of the shielding body or the position around the center for analysis and calculation. A series of related research has been conducted in this field. For example, an approximate statistical model of the internal field distribution of the complex cavity was deduced by the research team of Lehman and Hill of the Philip Laboratory of the US Air Force using the analytical method [17]. In addition, the team used the central limit theorem or the principle of maximum entropy for the complex cavity's internal field to study the complex cavity's internal field and established the distribution characteristics of the electromagnetic field and the statistical model of the electromagnetic field distribution and correlation coefficient [18,19]. Anlage [20,21] used the S-parameter matrix of the test port to evaluate the electromagnetic loss characteristics of the cavity and combined it with the Random Matrix Theory (RMT) to verify Hill's description of the statistical distribution law of the internal field of a complex cavity. The Federal Aviation Administration (FAA) concluded that the use of mode stirring could effectively test the electromagnetic attenuation coefficient of the aircraft cavity after testing the airworthiness of various types of aircraft [22].

Electromagnetic field analysis technology inside the cavity has gradually been phased in for practical use in laboratory research, and there are a few types of research on the irregular cavity. Meanwhile, the testing standards applicable for SE in different situations have been widely used and are given in [23–25], but sometimes these methods are not applicable for the SE testing of irregular cavity. In this paper, the effects of irregular structures on the cavity field distribution are analyzed, with the shielding effectiveness characterization and testing methods obtained based on the field distribution characteristics. The CST (3D EM Simulation Software) is used to simulate the field distribution in the irregular cavity.

2. Electromagnetic Topology Calculation Model of Shielding Effectiveness of Irregular Cavity

2.1. Electromagnetic Topology Modle

The physical model is shown of the irregular shielded cavity irradiated by plane waves in Figure 1. Based on the Robinson algorithm, the free space is equivalent to a transmission line with a characteristic impedance of Z_0, the front part and the back part of the cavity are equivalent to Z_{g1} and Z_{g2}, respectively, and both are rectangular waveguides, and the apertures and irregular structures are equal to two-port networks in the model.

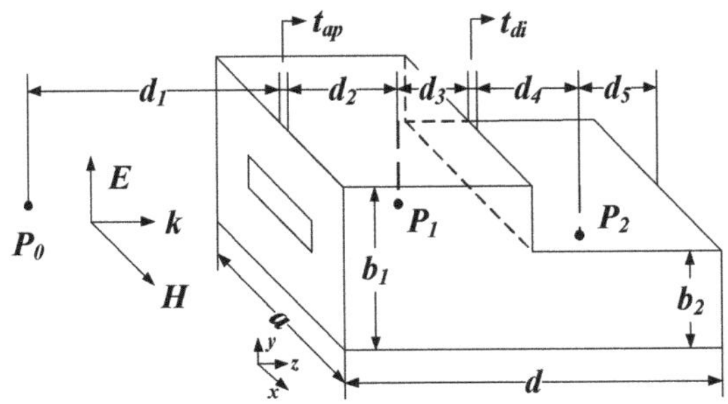

Figure 1. Physical model of irregular shielded cavity irradiated by plane waves.

Based on the electromagnetic topology theory, the above physical model is reconstructed as an equivalent transmission line network with tubes and nodes. The equivalent circuit diagram and the corresponding signal flow diagram are shown in Figures 2 and 3, respectively:

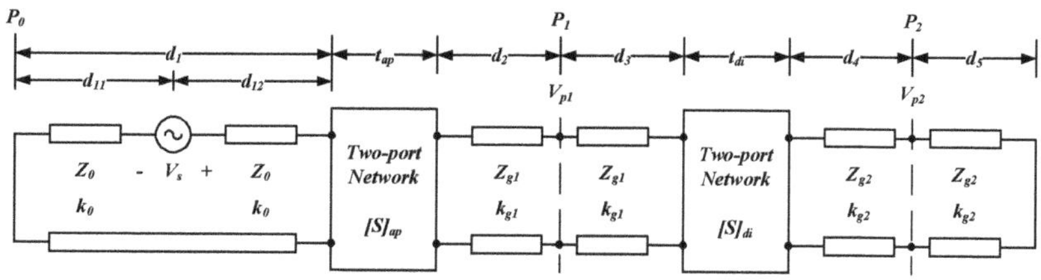

Figure 2. Equivalent circuit of an irregular shielded cavity irradiated by the plane waves.

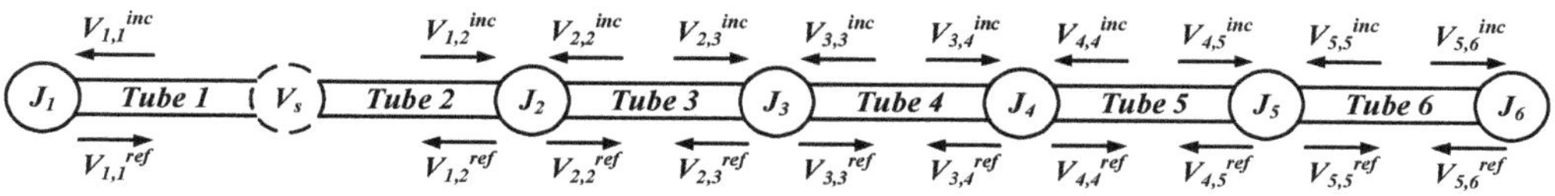

Figure 3. Signal flow diagram of an Irregular Shielded cavity irradiated by plane waves.

Based on the BLT (Baum–Liu–Tesche) matrix equation of the transmission line network, the BLT equation to solve the node voltage response is as follows:

$$V = [U + S] \cdot [\Gamma - S]^{-1} \cdot E \tag{1}$$

- V is the voltage vector.

$$V = [V_{1,1}, V_{1,2}, V_{2,2}, V_{2,3}, V_{3,3}, V_{3,4}, V_{4,4}, V_{4,5}, V_{5,5}, V_{5,6}]^T \tag{2}$$

The vector element $V_{i,j} = V_{i,j}{}^{inc} + V_{i,j}{}^{ref}$ is the full voltage at each node.

- U is the unit hypermatrix.
- S is the scattering matrix.

$$S = diag(\rho_1, S_2, S_3, S_4, S_5, \rho_6) \tag{3}$$

Because of the reflection coefficient of matching node J_1, $\rho_1 = 0$. J_6 is the reflection coefficient of the short-circuit node, and $\rho_6 = -1$. The scattering matrices of observation points $P_1(S_3)$ and $P_2(S_5)$ are $S_3 = S_5 = \begin{bmatrix} 0 & 1 \\ 1 & 0 \end{bmatrix}$.

The equivalent two ports of aperture structure node are T-type ones (Figure 4), and scattering matrix S_2 is:

$$S_2 = \begin{bmatrix} \frac{Y_1 - Y_2 - Y_a}{Y_1 + Y_2 + Y_a} & \frac{2Y_2}{Y_1 + Y_2 + Y_a} \\ \frac{2Y_1}{Y_1 + Y_2 + Y_a} & \frac{Y_2 - Y_1 - Y_a}{Y_1 + Y_2 + Y_a} \end{bmatrix} \tag{4}$$

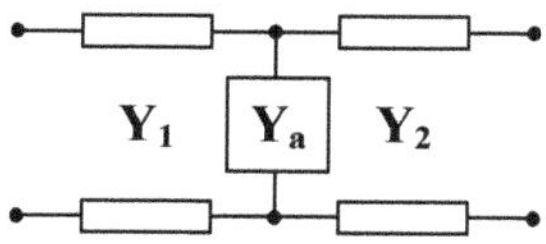

Figure 4. T-type equivalent two-port circuit.

The equivalent two-ports of the special-shaped structure node are Π-type ones (Figure 5), and scattering matrix S_4 is:

$$S_4 = \begin{bmatrix} \frac{Y_1{}^2 - 2Y_aY_b - Y_a{}^2}{(Y_1+Y_a)(Y_1+Y_a+2Y_b)} & \frac{2Y_1Y_b}{(Y_1+Y_a)(Y_1+Y_a+2Y_b)} \\ \frac{2Y_1Y_b}{(Y_1+Y_a)(Y_1+Y_a+2Y_b)} & \frac{Y_1{}^2 - 2Y_aY_b - Y_a{}^2}{(Y_1+Y_a)(Y_1+Y_a+2Y_b)} \end{bmatrix} \quad (5)$$

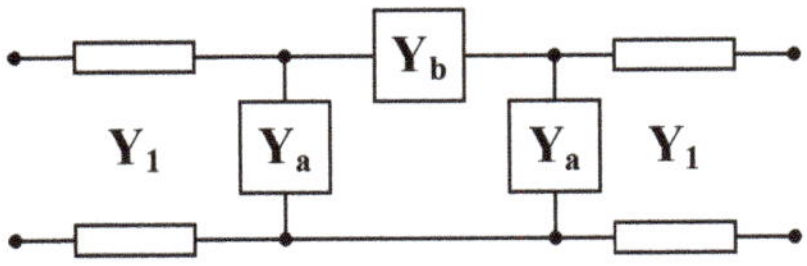

Figure 5. Π-type-equivalent two-port circuit.

- Γ is the propagation matrix.

$$\Gamma = diag(\Gamma_1, \Gamma_2, \Gamma_3, \Gamma_4, \Gamma_5) \quad (6)$$

In the formula, $\Gamma i = \begin{bmatrix} 0 & e^{\gamma_i d_i} \\ e^{\gamma_i d_i} & 0 \end{bmatrix}$.
Where $\gamma_1 = jk_0$, $\gamma_2 = \gamma_3 = jk_{g1}$, $\gamma_4 = \gamma_5 = jk_{g2}$, and d_i is the length of each tube.

- E is the source vector.

$$\mathrm{E} = \left[\frac{1}{2}V_s e^{\gamma_1 d_{11}} - \frac{1}{2}V_s e^{\gamma_1 d_{12}}\ 0\,0\,0\,0\,0\,0\,0\,0\right]^T \quad (7)$$

2.2. Irregular Cavity Classification

Based on the classification of irregular cavities in the electromagnetic topology theory, irregular cavities are typically divided into six types, as shown in Figure 6. When the electric field is perpendicular to the long side of the aperture, the most common irregular structures or a combination of the above six structures are covered.

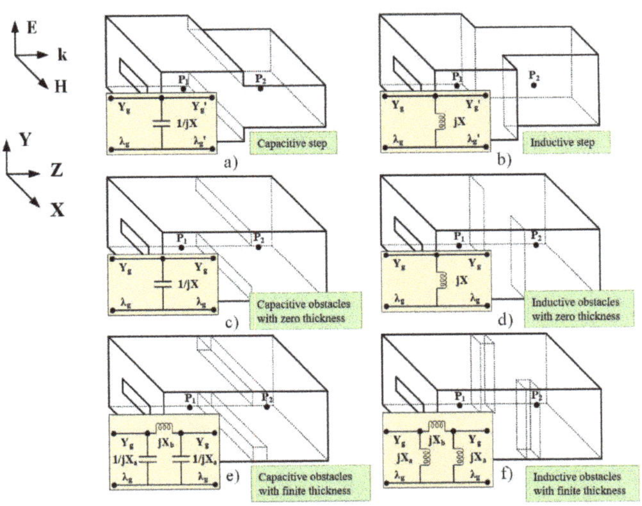

Figure 6. Six common types of irregular structures. (**a**) Capacitive step, (**b**) Inductive step, (**c**) Capacitive obstacles with zero thickness, (**d**) Inductive obstacles with zero thickness, (**e**) Capacitive obstacles with finite thickness, (**f**) Inductive obstacles with finite thickness.

The equivalent two-port types of six irregular structures are different. The structure of the step and the sheet is a T-type one, and the thick block structure is the Π-type one. In addition, the types of components in the equivalent two-port network of each irregular structure are also different. The parameters of equivalent circuit of irregular structure are as follows:

1. Capacitive step

In Figure 6a, the characteristic admittance of anterior chamber $Y_g = \sqrt{1-(\lambda/2a)^2}/Z_0$, and $Y'_g = Y_g b/b'$ is the characteristic admittance of the rear cavity.b' is the height of the rear cavity.

$$\frac{X}{Y_g} \approx \frac{2a}{\lambda_g}\left[ln\left(\frac{1-\alpha^2}{4\alpha}\right)\left(\frac{1+\alpha}{1-\alpha}\right)^{\frac{1}{2}(\alpha+\frac{1}{\alpha})}+\frac{2}{A}\right] \tag{8}$$

where, $\lambda_g = \lambda_0/\sqrt{1-(\lambda/2a)^2}$, $\alpha = b'/b$, $A = \left(\frac{1+\alpha}{1-\alpha}\right)^{2\alpha}\frac{1+\sqrt{1-(2b/\lambda_g)^2}}{1-\sqrt{1-(2b/\lambda_g)^2}} - \frac{1+3\alpha^2}{1-\alpha^2}$.

2. Inductive step

$$\begin{cases} \frac{X}{Y_g} \approx \frac{2a}{\lambda_g}2.33\alpha^2(1+1.56\alpha^2)(1+6.75\alpha^2 Q), \alpha \ll 1 \\ \frac{Y_g}{X} \approx \frac{\lambda_g}{2a}\frac{\beta^2(1+\beta)ln\frac{2}{\beta}}{1-\frac{\beta}{2}}\left(1-\frac{27}{8}\frac{Q+Q'}{1+8ln\frac{2}{\beta}}\right), \beta \ll 1 \end{cases} \tag{9}$$

where $\alpha = a'/a = 1-\beta$, $Q = 1-\sqrt{1-(2a/3\lambda)^2}$, $Q' = 1-\sqrt{1-(2a'/3\lambda)^2}$, $\lambda_g = \lambda/\sqrt{1-(\lambda/2a)^2}$, and ${\lambda_g}' = \lambda/\sqrt{1-(\lambda/2a')^2}$. a' is the width of the rear cavity.

3. Capacitive obstacles with a zero thickness value

$$\begin{cases} \frac{X}{Y_g} \approx \frac{4b}{\lambda_g}\left\{ln\left(\frac{2b}{\pi d}\right)+\frac{1}{6}\left(\frac{\pi d}{2b}\right)^2+\frac{1}{2}\left(\frac{b}{\lambda_g}\right)^2\left[1-\frac{1}{2}\left(\frac{\pi d}{2b}\right)^2\right]^4\right\}, \frac{d}{b} \ll 1 \\ \frac{X}{Y_g} \approx \frac{2b}{\lambda_g}\left[\left(\frac{\pi d'}{2b}\right)^2+\frac{1}{6}\left(\frac{\pi d'}{2b}\right)^4+\frac{3}{2}\left(\frac{b}{\lambda_g}\right)^2\left(\frac{\pi d'}{2b}\right)^4\right], \frac{d'}{b} \ll 1 \end{cases} \tag{10}$$

where d' is the height of two sheets, and d is the distance between two sheets.

4. Inductive obstacles with a zero thickness value

$$\begin{cases} \frac{X}{Y_g} \approx \frac{a}{\lambda_g}tan^2\left\{\frac{\pi d}{2a}\left[1+\frac{1}{6}\left(\frac{\pi d}{\lambda_0}\right)^2\right]\right\}, \frac{d}{a} \ll 1 \\ \frac{X}{Y_g} \approx \frac{a}{\lambda_g}cot^2\left\{\frac{\pi d'}{a}\left[1+\frac{2}{3}\left(\frac{\pi d'}{\lambda_0}\right)^2\right]\right\}, \frac{d'}{a} \ll 1 \end{cases} \tag{11}$$

where d' is the height of two sheets, and d is the distance between two sheets.

5. Capacitive obstacles with finite thickness

$$\begin{cases} \frac{X_a}{Y_g} = \frac{X}{Y_g} + \frac{b}{d}tan\frac{\pi l}{\lambda_g} \\ \frac{X_b}{Y_g} = \frac{b}{d}csc\frac{2\pi l}{\lambda_g} \end{cases} \tag{12}$$

where $\frac{X}{Y_g} = \frac{2b}{\lambda_g}\left[ln\frac{e}{\lambda_g}+\frac{\alpha^2}{3}+\frac{1}{2}\left(\frac{b}{\lambda_g}\right)^2(1-\alpha^2)^4\right]$, $\alpha = b'/b$, and b is the height of two blocks, and b' is the distance between two blocks. l is the thickness of two blocks.

6. Inductive obstacles with finite thickness

$$\begin{cases} \frac{X_a}{Y_g} = \frac{X}{Y_g} - \frac{b}{d}tan\frac{\pi l}{\lambda_g} \\ \frac{X_b}{Y_g} = \frac{b}{d}csc\frac{2\pi l}{\lambda_g} \end{cases} \tag{13}$$

where $\frac{X}{Y_g} = \frac{2b}{\lambda_g}\left[ln\frac{e}{\lambda_g} + \frac{\alpha^2}{3} + \frac{1}{2}\left(\frac{b}{\lambda_g}\right)^2\left(1-\alpha^2\right)^4\right]$, $\alpha = b'/b$, and b is the height of two blocks and b' is the distance between two blocks. l is the thickness of two blocks.

The shielding effectiveness of the observation points can be obtained by solving the BLT matrix equation of the node voltage response [26]. The calculation results are in good agreement with the simulation results when the frequency exceeds the first resonant frequency of the cavity. The solution takes only a few seconds, which is much faster than the simulation time required by the CST simulation. It also has a good prediction of the resonance frequency point of the irregular structure, and a relatively systematic analysis model of the shielding effectiveness of the irregular structure is presented.

3. Simulation Analysis of Field Distribution in Irregular Cavity

3.1. Simulation Parameters of Irregular Cavity Field Distribution

Since the field distribution in the cavity is affected by the size, shape, and other structural characteristics of the irregular cavity, the electromagnetic field distribution characteristics are mainly studied in the irregular cavity, with apertures in the frequency range of 20 MHz–1 GHz. In this process, the effects of the special-shaped structure on the distribution law of the electromagnetic field in the cavity are comprehensively analyzed.

The objects simulated are six primary, irregular cavities, with the size of the regular part set being 300 × 120 × 360 mm^3, the thickness of the cavity being 1 mm, and the size of the rectangular aperture being 40 × 20 mm^2.

The specific parameters of the six irregular cavities simulated by CST are shown in Table 1. The number of the steps or obstacles is two in each irregular cavity. d2 and d3 are the parameters shown in Figure 1.

Table 1. The parameters of six irregular structures.

Numbering	Type	Heteromorphic Structure Parameters
1	Capacitive step	300 × 20 × 160 mm^3
2	Inductive step	30 × 120 × 160 mm^3
3	Capacitive obstacles with a zero thickness value	300 × 20 × 1 mm^3, d2 +d3 = 180 mm
4	Inductive obstacles with a zero thickness value	25 × 120 × 1 mm^3, d2 +d3 = 180 mm
5	Capacitive obstacles with a finite thickness	300 × 10 × 40 mm^3, d2 +d3 = 200 mm
6	Inductive obstacles with a finite thickness	20 × 120 × 40 mm^3, d2 +d3 = 200 mm

3.2. Simulation Results of Field Distribution in the Irregular Cavity

The field distribution of six irregular cavities in the typical transmission mode Mode303 is simulated to study the effects of the irregular structure on the field distribution.

3.2.1. Basic Model Field Distribution

The basic model field in Mode303 is shown in Figure 7.

(a) (b) (c)

Figure 7. Field distribution of the basic model. (**a**) x section, (**b**) y section, and (**c**) z section.

The corresponding transmission modes are displayed on the cross-sections in all of the directions in the cavity when the frequency is in Mode = 303, with three half-wave transmission modes in the x-direction and the z-direction, which is in line with Mode303.

3.2.2. Field Distributions of Capacitive Step and Inductive Step

Mode303 field distribution diagrams of the two step structures are shown in Figures 8 and 9. Since the capacitive step changes the size in the y-direction, where there is no mode, the field distribution in the cavity shows no significant changes, while the inductive step changes the size in the x-direction, so the modes in the second half of the step structure are mixed, and the latter two modes are blended.

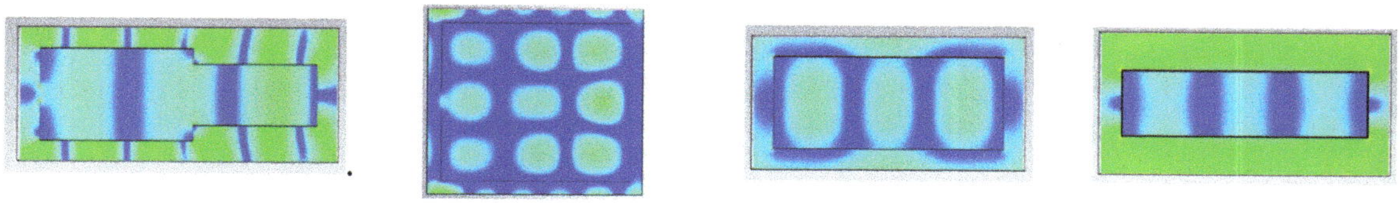

(a) (b) (c) (d)

Figure 8. Mode303 Field Distribution of Capacitive Step. (**a**) x section, (**b**) y section, (**c**) z section of the front part, and (**d**) z section of the back part.

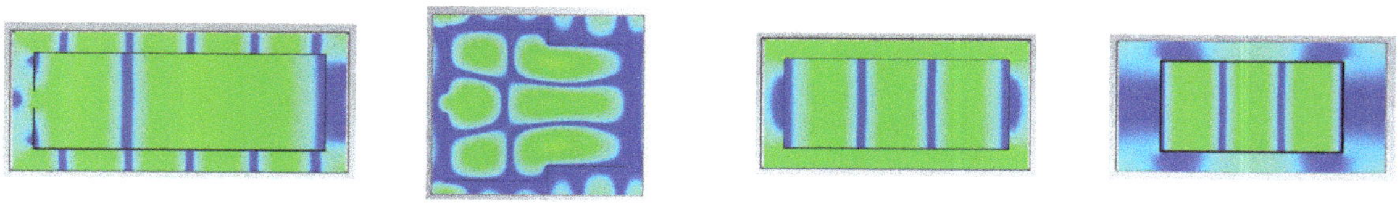

(a) (b) (c) (d)

Figure 9. Mode303 field distribution of inductive step. (**a**) x section, (**b**) y section, (**c**) z section of the front part, and (**d**) z section of the back part.

3.2.3. Field Distributions of Capacitive and Inductive obstacles with a Finite Thickness

Mode303 field distributions of the two thick block structures are shown in Figures 10 and 11.

The field distribution of capacitive obstacles with a finite thickness is more in line with the basic model than the inductive thick block cavity is, which is slightly different and can be seen more clearly from the y section view. In contrast, the distribution of the modes becomes blurred due to the cross-coupling of the modes in the x-direction and z-directions.

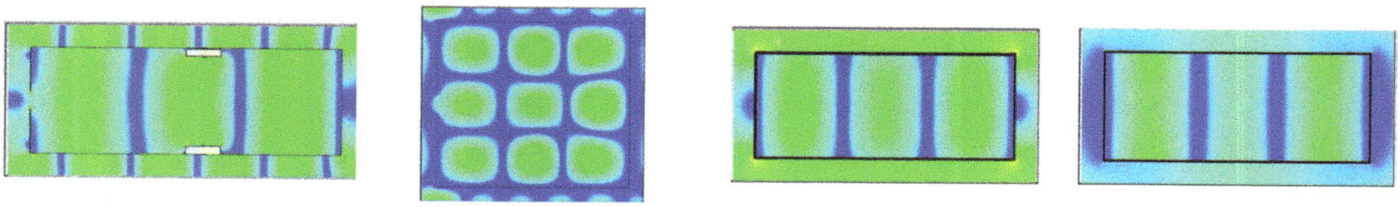

(a) (b) (c) (d)

Figure 10. Mode303 field distribution of capacitive obstacles with a finite thickness. (**a**) x section, (**b**) y section, (**c**) z section of the front part, and (**d**) z section of the back part.

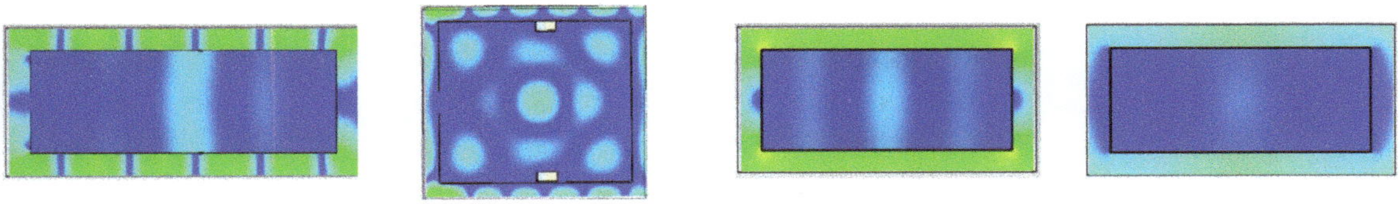

(a) (b) (c) (d)

Figure 11. Mode303 field distribution of inductive obstacles with a finite thickness. (**a**) x section, (**b**) y section, (**c**) z section of the front part, and (**d**) z section of the back part.

3.2.4. Field Distributions of Capacitive and Inductive Obstacles with Zero Thickness

Mode303 field distributions of the two obstacles with a zero thickness value are shown in Figures 12 and 13. The field distribution of the capacitive obstacles with a zero thickness value is in line with the basic model. The changes to both of them in the cavity are relatively consistent with the cross-coupling.

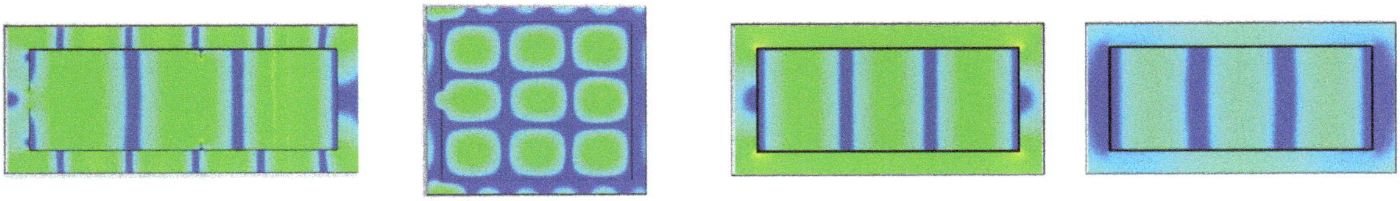

(a) (b) (c) (d)

Figure 12. Mode303 field distribution of capacitive obstacles with a zero thickness value. (**a**) x section, (**b**) y section, (**c**) z section of the front part, and (**d**) z section of the back part.

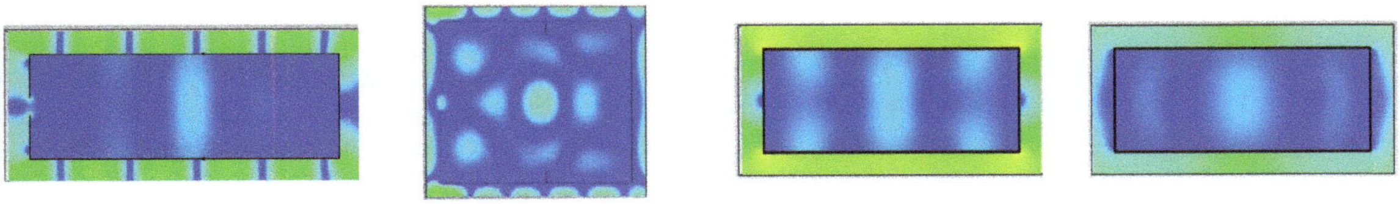

(a) (b) (c) (d)

Figure 13. Mode303 field distribution of inductive obstacles with a zero thickness. (**a**) x section, (**b**) y section, (**c**) z section of the front part, and (**d**) z section of the back part.

The field distribution of the capacitive obstacles with a zero thickness value is aligned well with the basic model with similar sections, and it has only a small effect on the field distribution. In contrast, there are some changes in the field distribution of the inductive obstacles with a zero thickness value on the y section, mainly reflected in more independent modes and fewer changes in the overall field distribution due to the intervention of the irregular structures that separate the previously connected modes.

3.3. Analysis of Field Distribution in Special-Shaped Cavity

The field distribution of the irregular cavity has been simulated and analyzed before. The typical resonance frequency has been selected for simulation research, mainly for the field distribution of six basic, irregular cavities. However, this analysis concerns on the effects of the irregular structures on the field distribution and the selection of observation points in the shielding effectiveness testing method.

1. The effects of irregular structures.

The inductive, irregular structure has a more significant impact on the field distribution than the capacitive one does. As the electric field polarization direction is the height direction in the simulated plane wave, the field distribution of the cavity may be more sensitive to the size or structure changes perpendicular to the plane wave polarization direction.

2. Selection of observation points.

The overall change in the intensity of the electric field and the position where the higher field intensity is likely to occur are analyzed to select the appropriate points for observation based on the test requirements on the ground and the field distribution characteristics of the irregular cavity.

For irregular cavities, the overall distribution characteristics are in line with those of the regular cavities only, with the following differences:

- Irregular cavities mean that the distributions of the front and back parts are different, with the cavity of the step structure being the most obvious part. As the step structure changes the size of the front and back parts of the cavity, and the change in the size of the capacitive step is the height direction, the mode transmission has only a small effect on the overall field distribution since there is no mode transmission in the height direction of the back part of the cavity in the Mode303. However, there are changes in the length direction of the back part of the cavity in the inductive steps, resulting in the coupling of the modes in the front and back parts of cavities.
- The regular cavity is similar to the irregular cavity of the thick block type on the whole, which is mainly reflected in the cross-coupling of the inductive thick block appearance mode.
- As there is a small volume in the cavity that the thin plate takes up, the capacitive thin plate has only a small effect on the field distribution of the cavity. In contrast, the inductive thin plate has more significant effects, which is similar to the inductive thick block.

4. Characterization and Test Method of Shielding Effectiveness of Irregular Cavity

4.1. Characterization of Shielding Effectiveness of Irregular Cavity

Shielded cavities have a wide range of types and applications, and testing standards applicable in different situations can be found for different cavities [24–26]. Among these standards, IEEE 299-2006 and IEEE 299.1-2013 are the most widely used ones, and the shielding effectiveness test methods for cabins that are above 2 m and 0.1–2 m cabins are given, respectively.

The number of modes and mode distribution of the field distribution inside the cavity change with frequency based on the research on the field distribution of the irregular cavity. The field strength obtained from relatively fewer observation points may not be the maximum field strength on the ground in the cavity. Therefore, it is necessary to reconsider the position of the field strength monitoring points applicable to the shielding effectiveness of the irregular cavities.

On the one hand, the field distribution in the shielding cavity is affected by the irregular structure. In the low-frequency band with few modes in the cavity, the appropriate observation node position can be selected during the test based on the mode distribution. Generally, the center position of the cavity can be selected. Since this position is at the center of the length, height, and width directions, modes are more likely to appear at the center of the length and height directions or keep a fixed distance from the wall in the width direction, such as 0.3 m, which is given in the standard. In the high-frequency band with more modes in the height direction, the maximum field strength in the cavity may not be measured in a precise way by the original observation points, a particular movement in the height direction is taken into account, or two center positions can be selected that are bisected in the height direction.

On the other hand, as the internal space of the shielding cavity is limited, the test antenna's volume needs to be considered from the actual testing perspective. It may not be possible to increase the observation points, which partly explains why the standard only provides observation points at a fixed distance from the wall, as one or two observation points need be added in a specific direction of a particular frequency band in the large cavity.

4.2. Test Method for Shielding Effectiveness of Irregular Cavity

For the irregular cavity, the antenna size cannot be measured by the regular test method, including the biconical antenna between 20 MHz and 100 MHz specified in the standard and half-wave dipole antennas or logarithmic periodic antennas between 100 MHz and 1 GHz because of their large size. In addition, as the polarization direction of the antenna needs to be considered, the space for the antenna must be provided in both of the directions, which is not applicable in the cavity with limited space because the coupling between the antenna and the wall will make the measured results inaccurate.

In the frequency band from 20 MHz to 1 GHz, the laboratory's self-developed receiving spherical dual antenna is used for the shielding effectiveness test of the irregular cavity. The plan for the test is given based on the shielding effectiveness characterization and selecting the location and number of observation points, as shown in Figures 12 and 13. The transmitting antenna is the biconical antenna (20 MHz~100 MHz) or logarithmic period antenna (100 MHz~1 GHz).

The layout of a certain observation point is shown in Figures 14 and 15. The position and number of observation points can be selected based on the analysis above of the shielding effectiveness of the irregular cavity in Table 2.

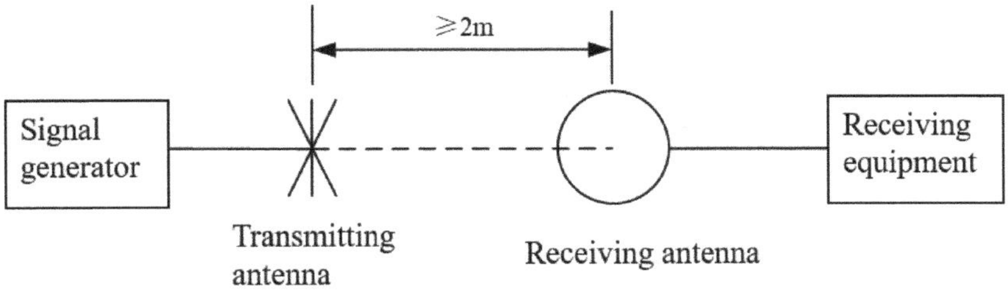

Figure 14. Reference test placement.

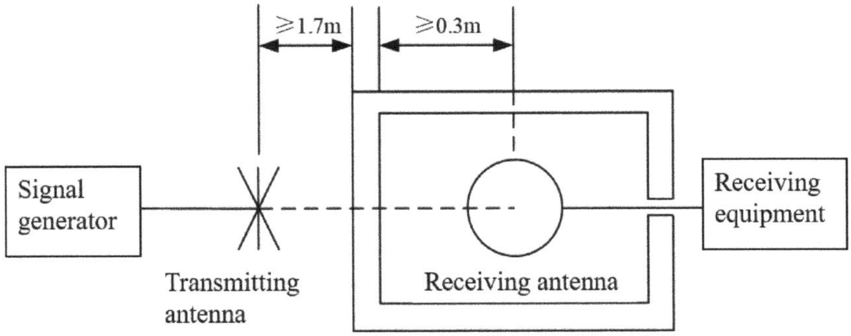

Figure 15. Shielding test arrangement.

Table 2. Location and number of observation points.

Test Frequency Band	Location and Number of Observation Points
20 MHz~0.8f_r	Single point: Cavity center position
0.8f_r~1 GHz	Multi-point: the center position of the x-direction, the y-direction or the center position of the multi-segment division, and the center position of the multi-segment division in the z-direction

In the table, f_r is the first resonance frequency of the cavity. At 20 MHz~0.8 f_r, no transmission mode appears in the cavity, so the center position of the cavity can be used during the test; at 0.8 f_r~1 GHz, the mode is formed in the cavity. During the test, selecting multiple equally divided center positions as observation points in all of the directions is necessary.

The antenna polarization direction is arranged based on the test requirements. The test is generally carried out at horizontal and vertical polarizations if no particular need exists.

5. Measurement for Shielding Effectiveness of Irregular Cavity

The self-developed, spherical dipole receiving antenna is used to test the shielding effectiveness of the irregular cavity based on the proposed test method for the shielding effectiveness of the irregular cavity.

5.1. Test Parameters

The test object is selected as the existing aluminum cavity in the laboratory, with a size of 470 mm × 400 mm × 610 mm; the test irregular structure is chosen as an aluminum tube, with a size of 100 mm × 100 mm × 100 mm, which is placed in the middle of the bottom of the cavity as a step-like structure; the frequency band is 20 MHz~1 GHz; the signal source is placed on the outside during the test.

5.2. Test Layout

5.2.1. Reference Test Arrangement

The actual test setup for the test (use the test setup of the horizontal polarization of the biconical antenna in the 20 MHz~100 MHz frequency band as an example) is shown in Figure 16. The layout in the 100 MHz~1 GHz frequency band is the same, substituting the biconical antenna for a periodic log antenna.

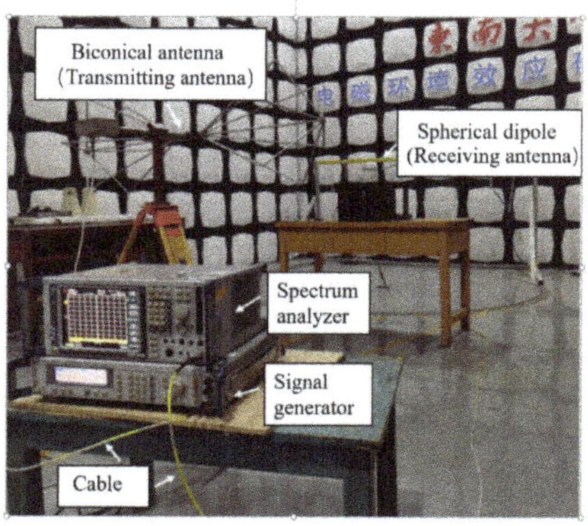

Figure 16. The actual layout of the reference test of the cavity shielding effectiveness (horizontal polarization of the biconical antenna).

5.2.2. Shield Test Arrangement

The internal and external layout of the cavity (taking vertical polarization as an example) during the shielding test of the cavity shielding effectiveness is shown in Figure 17, where the aluminum cube is placed in the middle of the bottom of the cavity. The overall setup of the test system is similar to that of the reference test shown in Figure 16.

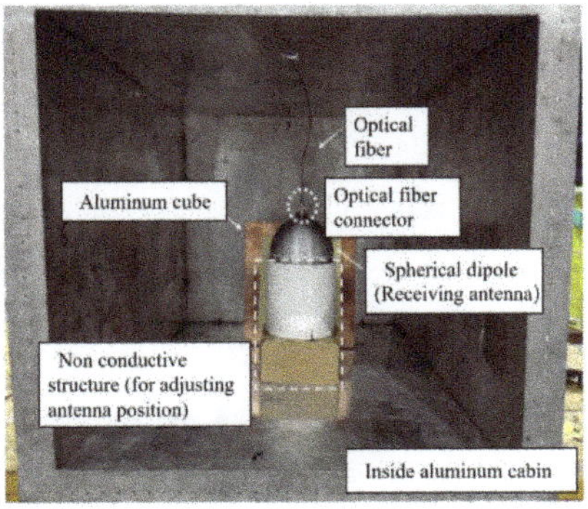

(a)

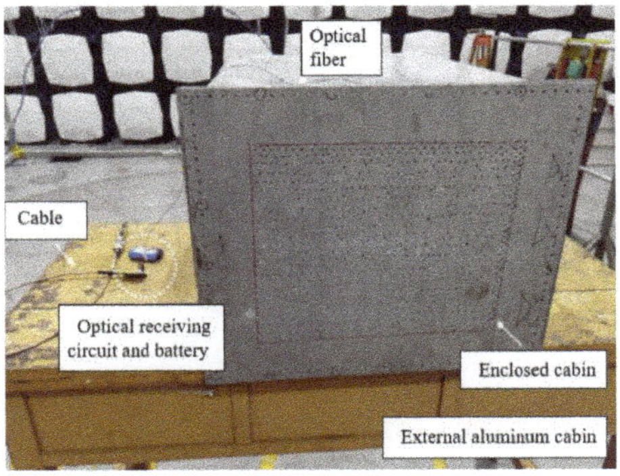

(b)

Figure 17. The internal and external layout of cavity shielding effectiveness (vertical antenna polarization). (**a**) Layout inside the cavity; (**b**) external arrangement of the cavity.

5.3. Test Results

During the reference test, the output power of the signal source is set as −10 dBm, and a scatter test is carried out within 20 MHz~1 GHz.

During the shielding test, the output power of the signal source remains unchanged, and the first resonant frequency f_r is 550.4 MHz based on the size of the tested cavity: 470 mm × 400 mm × 610 mm. Since the shape of the aperture of the tested cavity is circular, and the length size is similar to the height, the patterns in the two directions that may appear in the interior are unlikely to have different parities.

The following observation points are selected based on Table 2, and a schematic diagram of the location of observation points is shown in Figure 16. For the observation points between 20 MHz and 440 MHz (0.8 f_r), a single point is selected, which is the No. 1 point in the center of the cavity. For observation points between 440 MHz (0.8 f_r) and 1 GHz, single or multiple points can be selected for testing, where single points at the center position of cavity No. 1 and multi-points of the bisecting point in the length direction and height direction are chosen for comparative testing. In addition, confined by the aluminum cube structure, the width direction is only selected as the center position, with four observation points from No. 2 to No. 5. A schematic diagram of the location of observation points is shown in Figure 18.

Figure 18. Schematic diagram of observation points location.

The shielding test results of the tested cavity's shielding effectiveness are shown in Tables 3 and 4. While Table 3 shows the test results of shielding effectiveness at the No. 1 point in the center of the cavity in the frequency band of 20 MHz~1 GHz, Table 4 shows the comparison result of single-point and multi-point test data in the frequency band of 440 MHz (0.8 f_r)~1 GHz.

Table 3. Shielding effectiveness of point No. 1 in the center of the test cavity observation points.

Figure	Type of Transmitting Antenna	Polarization Direction	Shielding Effectiveness/dB
30	Biconical antenna	Horizontal	38.08
		Vertical	34.34
50	Biconical antenna	Horizontal	34.5
		Vertical	35.85
80	Biconical antenna	Horizontal	32.97
		Vertical	33.11
100	Biconical antenna	Horizontal	43.06
		Vertical	42.82
100	Periodic log antenna	Horizontal	46.23
		Vertical	42.63
200	Periodic log antenna	Horizontal	34.14
		Vertical	31.92
300	Periodic log antenna	Horizontal	26.54
		Vertical	31.49
500	Periodic log antenna	Horizontal	25.07
		Vertical	28.24
800	Periodic log antenna	Horizontal	18.21
		Vertical	19.2
1000	Periodic log antenna	Horizontal	14.26
		Vertical	16.68

Table 4. Shielding effectiveness of test chamber observation points from No.1 to No.5.

Frequency/ MHz	Polarization Direction	No.1 SE/dB	No.2 SE/dB	No.3 SE/dB	No.4 SE/dB	No.5 SE/dB
500	Horizontal	25.07	24.79	26.59	24.83	**24.25**
	Vertical	28.24	28.72	**27.19**	28.78	27.7
800	Horizontal	18.21	**17.88**	18.12	18.89	18.03
	Vertical	19.2	20.85	**18.65**	20.03	19.07
1000	Horizontal	**14.26**	14.51	14.29	15.08	14.85
	Vertical	16.68	**16.64**	17.44	17.04	17.25

It can be seen from Table 3 that the single observation points test of the cavity shielding effectiveness is carried out smoothly. The results of different observation points in the 440 MHz (0.8 f_r)~1 GHz frequency band in Table 4 can be analyzed. The frequency points and the minimum shielding effectiveness observation points in the same antenna polarization directions are marked in Table 4.

Based on the test results, it can be found that when the frequency band is high, the modes in the cavity gradually increase. When only the center position of the cavity is used for testing, other locations with more immense field strengths are likely to be ignored, resulting in a more significant shielding effect.

When the shielding effectiveness of the irregular cavity is tested, more test points can undoubtedly improve the accuracy. Still, the time and labor that it takes will be significantly increased. If the characteristics of the body in the frequency band to be measured are analyzed in advance, selecting a small, but appropriate, number of observation points can effectively improve the validity of the test results and can significantly reduce the test time. In addition, the test accuracy can be effectively improved through the selection of multi-point observation points, which serve as a reference to each other when one is testing different observation points, avoiding the occurrence of invalid points with significant data deviations, and reducing the possibility of re-testing due to problems cropping up in operation.

6. Conclusions

In this paper, the method of selecting monitoring points and measuring the shielding effectiveness based on the field distribution is proposed. An electromagnetic topology model to analyze the shielding effectiveness is built first, which has a good prediction of the resonance frequency point of the irregular structure. Then, the appropriate points are selected for observation based on the test requirements on the ground and the field distribution characteristics of the irregular cavity. In the modeless range (20 MHz~$0.8f_r$), the center position of the cavity can be used as the observation points during the test; in the under-mode range ($0.8f_r$~1 GHz), multiple equally divided center positions are selected as observation points in all directions. The shielding effectiveness of irregular cavities can be tested accurately using this method. If the characteristics of the body in the frequency band to be measured are analyzed in advance, selecting a small, but appropriate, number of observation points can effectively improve the validity of the test results and can significantly reduce the test time.

This method can be used to test the shielding effectiveness of the irregular cavity, and the effectiveness of the method is verified in this paper. The following work was conducted to further simplify this testing method.

Author Contributions: Conceptualization, J.R.; Methodology, J.R.; Software, Y.P.; Validation, J.R. and T.Z.; Formal analysis, Y.P.; Investigation, Z.Z.; Data curation, Y.P.; Writing—original draft, J.R. and Y.P.; Writing—review & editing, J.R. and Z.Z.; Project administration, Z.Z.; Funding acquisition, J.R. and T.Z. All authors have read and agreed to the published version of the manuscript.

Funding: This work was supported in part by the Aero-Science Fund of China, under Grant 201926069001, and the National Defense Basic Research Program of China, under Grant No. JCKYS2020 DC404.

Data Availability Statement: The data used to support the findings of this study are included within the article.

Acknowledgments: Detailed steps and comparison of the electromagnetic topology calculation model of the shielding effectiveness of the irregular cavity were presented at the 2021 IEEE 5th Information Technology, Networking, Electronic and Automation Control Conference (ITNEC) [26].

Conflicts of Interest: The authors declare no conflict of interest.

References

1. Jóskiewicz, Z.; Janukiewicz, J. Experimental Study of the Shielding Effectiveness Performance Degradation for a Shielding Material Used in Protective Storage Pouch. In Proceedings of the 2020 International Symposium on Electromagnetic Compatibility-EMC EUROPE, Rome, Italy, 23–25 September 2020; pp. 1–4.
2. Li, H.; Bao, X.; Zhou, H. Electromagnetic Pulse Propagation over Large Area Simulation through Massively Parallel Adaptive Mesh Refinement FDTD. In Proceedings of the 2019 International Symposium on Electromagnetic Compatibility-EMC EUROPE, Barcelona, Spain, 2–6 September 2019.
3. Xiao, L.; Wang, X.H.; Wang, B.Z.; Zheng, G.; Chen, P. An Efficient Hybrid Method of Iterative MoM-PO and Equivalent Dipole-Moment for Scattering From Electrically Large Objects. *IEEE Antennas Wirel. Propag. Lett.* **2017**, *16*, 1723–1726. [CrossRef]
4. Geng, L.; Liu, X.; He, H. Analytical formulation for the coupling coefficient of equivalent impedance to calculated shielding effectiveness by TLM. In Proceedings of the 2017 IEEE 5th International Symposium on Electromagnetic Compatibility (EMC-Beijing), Beijing, China, 28–31 October 2017.
5. Cruciani, S.; Feliziani, M.; Maradei, F. Prediction of shielding effectiveness in graphene enclosures by FEM-INBC method. In Proceedings of the 2015 Asia-Pacific Symposium on Electromagnetic Compatibility, Taipei, Taiwan, 26–29 May 2015.
6. Wang, C.F. Development of Higher-Order DG-FETD Tool for Efficient Modeling of Complex Electromagnetic Structures. In Proceedings of the 2019 IEEE International Symposium on Antennas and Propagation and USNC-URSI Radio Science Meeting, Atlanta, GA, USA, 7–12 July 2019.
7. Li, M.; Liu, B.; Wei, Q.; Yu, K.; Wang, C.; Zeng, X. Time-Frequency Analysis on Earth Fault Signal Response Based on the BLT Equation for MultiBranch Distribution Networks. In Proceedings of the 2020 12th IEEE PES Asia-Pacific Power and Energy Engineering Conference (APPEEC), Nanjing, China, 20–23 September 2020; pp. 1–5.
8. Choo, J.; Choo, J.; Kim, Y.H. Shielding Effectiveness of Open Cabinet Containing Digital Modules Using Ferrite Sheet. *IEEE Trans. Magn.* **2017**, *53*, 2900609. [CrossRef]

9. Konefal, T.; Dawson, J.F.; Marvin, A.C.; Robinson, M.P.; Porter, S.J. A fast multiple mode intermediate level circuit model for the prediction of shielding effectiveness of a rectangular box containing a rectangular aperture. *IEEE Trans. Electromagn. Compat.* **2005**, *47*, 678–691. [CrossRef]
10. Robinson, M.P.; Benson, T.M.; Christopoulos, C.; Dawson, J.F.; Ganley, M.D.; Marvin, A.C.; Porter, S.J.; Thomas, D.W. Analytical formulation for the shielding effectiveness of enclosures with apertures. *IEEE Trans. Electromagn. Compat.* **1998**, *40*, 240–248. [CrossRef]
11. Solin, J.R. Formula for the field excited in a rectangular cavity with a small aperture. *IEEE Trans. Electromagn. Compat.* **2011**, *53*, 82–90. [CrossRef]
12. Solin, J.R. Formula for the field excited in a rectangular cavity with an electrically large aperture. *IEEE Trans. Electromagn. Compat.* **2012**, *54*, 188–192. [CrossRef]
13. Lee, J.G.; Eom, H.J.; Kim, B.W.; Park, H.H. Shielding effectiveness of enclosure with thick multiple apertures. *Microw. Opt. Technol. Lett.* **2001**, *29*, 178–181. [CrossRef]
14. Wallyn, W.; Zutter, D.D.; Laermans, E. Fast shielding effectiveness prediction for realistic rectangular enclosures. *IEEE Trans. Electromagn. Compat.* **2003**, *45*, 639–643. [CrossRef]
15. Park, H.H.; Eom, H.J. Electromagnetic penetration into a rectangular cavity with multiple rectangular apertures in a conducting plane. *IEEE Trans. Electromagn. Compat.* **2000**, *42*, 303–307. [CrossRef]
16. Konefal, T.; Dawson, J.F.; Marvin, A.C.; Robinson, M.P.; Porter, S.J. A fast circuit model description of the shielding effectiveness of a box with imperfect gaskets or apertures covered by thin resistive sheet coatings. *IEEE Trans. Electromagn. Compat.* **2006**, *48*, 134–144. [CrossRef]
17. Lehman, T.H.; Miller, E.K. The elementary statistical properties of electromagnetic fields in complex cavities. In Proceedings of the ICAP 91, Seventh International Conference on Antennas and Propagation, York, UK, 15–18 April 1991; Volume 2, pp. 938–941.
18. Hill, D.A. Plane wave integral representation for fields in reverberation chambers. *IEEE Trans. Electromagn. Compat.* **1998**, *40*, 209–217. [CrossRef]
19. Hill, D.A.; Ladbury, J.M. Spatial-correlation functions of fields and energy density in a reverberation chamber. *IEEE Trans. Electromagn. Compat.* **2002**, *44*, 95–101. [CrossRef]
20. Gradoni, G.; Antonsen, T.M.; Ott, E.; Anlage, S.M. Random coupling model for the radiation of statistical sources inside cavities. In Proceedings of the 2016 46th European Microwave Conference, London, UK, 4–6 October 2016.
21. Gil, J.G.; Drikas, Z.B.; Andreadis, T.D.; Anlage, S.M. Prediction of Induced Voltages on Ports in Complex, Three-Dimensional Enclosures With Apertures, Using the Random Coupling Model. *IEEE Trans. Electromagn. Compat.* **2016**, *58*, 1535–1540. [CrossRef]
22. *The Certification of Aircraft Electrical and Electronic Systems for Operation in the High-Intensity Radiated Fields (HIRF) Environment*; AA AC 20-158A (05/14); Federal Aviation Administration: Washington, DC, USA, 2014.
23. *IEEE Std 299-2006*; IEEE Standard Method for Measuring the Effectiveness of Electromagnetic Shielding Enclosures. IEEE: Piscataway, NJ, USA, 2006.
24. *IEEE Std 299.1-2013*; IEEE Standard Method for Measuring the Shielding Effectiveness of Enclosures and Boxes Having All Dimensions between 0.1 m and 2 m. IEEE: Piscataway, NJ, USA, 2013.
25. *IEC TS 61587-3-2013*; Mechanical Structures for Electronic Equipment–Test for IEC 60917 and IEC 60297–Part 3: Electromagnetic Shielding Performance Tests for Cabinets, Racks and Subracks. IEC: Geneva, Switzerland, 2013.
26. Pan, Y.; Ren, J.; Zhou, Z.; Zhang, T. Research on Calculation Method of Shielding Effectiveness of Irregular Cavity. In Proceedings of the 2021 IEEE 5th Information Technology, Networking, Electronic and Automation Control Conference (ITNEC), Xi'an, China, 15–17 October 2021.

electronics

Communication

A Novel Tuning Fork-Shaped Tri-Band Planar Antenna for Wireless Applications

Qiwei Li [1,2], Jinyong Fang [2,*], Jun Ding [1,*], Wen Cao [3], Jing Sun [2], Chenjiang Guo [1] and Tao Liu [4]

1 School of Electronics and Information, Northwestern Polytechnical University, Xi'an 710129, China
2 China Academy of Space Technology (Xi'an), Xi'an 710100, China
3 School of Electronics and Control Engineering, Chang'an University, Xi'an 710064, China
4 School of Electronic Engineering, Xidian University, Xi'an 710071, China
* Correspondence: fangjy111@163.com (J.F.); dingjun@nwpu.edu.cn (J.D.)

Abstract: A novel tuning fork-shaped tri-band planar antenna (NTTPA) for the LTE 2.3/3.8-GHz band, WLAN 2.4/5.2/5.8-GHz band, and WiMax 2.5/3.5/5.5-GHz band is presented in this letter. By introducing an asymmetrical turning fork-shaped patch and an inverted L-shaped patch, three notched bands can be generated to form a triple-band operation. The antenna is fabricated on an FR4 board and excited by an SMA connector using a microstrip line. The antenna structure is simple and has a compact size of 45 mm $\times$ 40 mm. The measured operating frequency covers 2.2–2.63, 2.73–3.8, and 5.13–6.3 GHz, and the percentage bandwidth is close to 53.3% (S11 < −9.8 dB from 2.2 to 3.8 GHz) and 20.5% (S11 < −10 dB from 5.13 to 6.3 GHz). The calculated and experimental results suggest that the proposed antenna is one of the best candidates for wireless communication systems in terms of multi operating bands, broad percentage bandwidth (BW), compactness, stable radiation pattern, easy processing, and low cost.

Keywords: multiband planar antenna; wireless communication; impedance bandwidth (BW); return loss (RL); radiation pattern

1. Introduction

In recent years, wireless communication technology has made great achievements and has been widely used in all kinds of electronic equipment. The antenna, which plays a key role in wireless communication systems, should be capable of operating at multi-frequencies simultaneously. The microstrip antenna is an attractive candidate for wireless applications, owing to its characteristics of low profile, low cost, easy processing, and manufacturing [1,2].

Over the past decades, numerous microstrip patch antennas with multi operating frequencies have been reported on for wireless communication, particularly for Long Term Evolution (LTE 2.3/3.8-GHz), wireless local area networks (WLAN 2.4/5.2/5.8-GHz), and worldwide interoperability for microwave access (WiMAX 2.5/3.5/5.5-GHz). These include asymmetric m-shaped antennas with vias [3], dielectric-loaded monopoles with shorted loops [4], proximity-coupled dual-substrate antennas with corner-truncated rectangular patches and defected ground planes [5], dumb-bell-shaped defected structure monopoles [6], open slot antennas [7], uniplanar dipole complementary capacitively loaded loop (CCLL) slots [8], rectangular ring monopoles with fork-shaped strips [9], dipoles with two electrical shorts and reflecting ground planes [10], D-shaped monopoles [11], patch antennas with inner patches and outer rings [12], and multi feed antennas [13–16] covering different wireless frequencies. The majority of the designs mentioned above only cover parts of the bands of LET, WLAN, and WiMax [3–7,9,12–25]. The designs are in the references. The examples in [8,10,11] can basically meet all frequency band requirements, but these antennas have a relatively large size. Hence, it is a huge challenge to design a multiband microstrip antenna not only covering the all the LTE (2.3–2.4, 3.6–3.8 GHz),

Citation: Li, Q.; Fang, J.; Ding, J.; Cao, W.; Sun, J.; Guo, C.; Liu, T. A Novel Tuning Fork-Shaped Tri-Band Planar Antenna for Wireless Applications. *Electronics* **2023**, *12*, 1081. https://doi.org/10.3390/electronics12051081

Academic Editor: Reza K. Amineh

Received: 23 November 2022
Revised: 8 February 2023
Accepted: 17 February 2023
Published: 22 February 2023

WLAN (2.4–2.484, 5.15–5.35, 5.725–5.825 GHz), and WiMAX (2.5–2.69, 3.4–3.69, 5.25–5.85 GHz) bands but also maintaining a compact size.

This letter presents a novel tuning fork-shaped tri-band planar antenna (NTTPA) for LTE (2.3/3.8-GHz), WLAN (2.4/5.2/5.8-GHz), and WiMAX (2.5/3.5/5.5-GHz) applications. The proposed antenna is composed of an asymmetrical turning fork-shaped patch and an inverted L-shaped patch connected to the middle of an rectangular ground, which can achieve a good performance of multi operation bands, enhanced percentage bandwidth, stable radiation patterns at different operating bands, and low cost. The proposed antenna has a compact size of 45 mm × 40 mm and is much smaller than [8,10,11]. That is to say, the antenna mentioned in Table 1 and in this paper can keep good performance in three frequency bands while realizing the miniaturization of size. The details of the design, parameter studies, and calculated and experimental results such as RL characteristics, surface current distributions, radiation patterns, and gains are also discussed and presented in the following chapters.

Table 1. Comparison of proposed NTTPA with previous antennas.

Ref.	Antenna Reponse	Dimensions ($\lambda_0 \times \lambda_0$)	Working Bands (GHz)	Radiation Effciency (%)	Gain (dBi)
[3]	Triple-band	$0.25\lambda_0 \times 0.53\lambda_0$	2.38–2.53 GHz 3.08–3.80 GHz 5.00–6.90 GHz	88% 85% 75%	1.1–1.5 dBi 4.6–5.6 dBi 2.0–3.6 dBi
[8]	Triple-band	$0.17\lambda_0 \times 0.32\lambda_0$	2.10–2.49 GHz 3.22–4.30 GHz 4.89–6.12 GHz	72% 69% 80%	2.7–3.2 dBi 3.1–3.5 dBi 2.8–3.3 dBi
[11]	Triple-band	$0.23\lambda_0 \times 0.27\lambda_0$	2.29–2.88 GHz 3.26–3.88 GHz 4.17–6.07 GHz	85% 84% 83%	3.8–4.4 dBi 4.0–4.7 dBi 1.9–3.5 dBi
[15]	Triple-band	$0.21\lambda_0 \times 0.25\lambda_0$	2.33–2.55 GHz 3.00–3.88 GHz 5.15–5.90 GHz	62% 70% 79%	1.1–1.4 dBi 2.4–3.5 dBi 2.5–3.3 dBi
[26]	Triple-band	$0.2\lambda_0 \times 0.25\lambda_0$	2.50–2.71 GHz 3.37–3.63 GHz 5.20–5.85 GHz	98% 96% 94%	1.3–2.5 dBi 1.4–2.4 dBi 1.3–2.6 dBi
[27]	Triple-band	$0.33\lambda_0 \times 0.37\lambda_0$	2.47–2.65 GHz 3.27–3.63 GHz 5.20–5.83 GHz	/ / /	1.8–2.4 dBi 1.8–2.4 dBi 1.7–2.4 dBi
Pre.	Triple-band	$0.33\lambda_0 \times 0.29\lambda_0$	2.20–2.63 GHz 2.73–3.80 GHz 5.13–6.30 GHz	80% 85% 90%	0.9–2.0 dBi 2.0–3.8 dBi 4.8–5.7 dBi

2. Antenna Topology

The geometry of the proposed novel tuning fork-shaped tri-band planar antenna (NTTPA), including the top view, side view, and bottom view, is depicted in Figure 1. The proposed NTTPA is fabricated on a single-layer FR4 substrate with a height of h = 1.6 mm, a relative dielectric constant of 4.4, and a loss tangent of 0.02. As shown in Figure 1, the proposed NTTPA is composed of an asymmetrical tuning fork-shaped patch and an inverted L-shaped monopole combined with a rectangular ground plane fed by an SMA connector using a microstrip line. Table 2 displays the overall dimensions of the proposed antenna. To facilitate processing, all line widths are designed to be the same W = 3 mm.

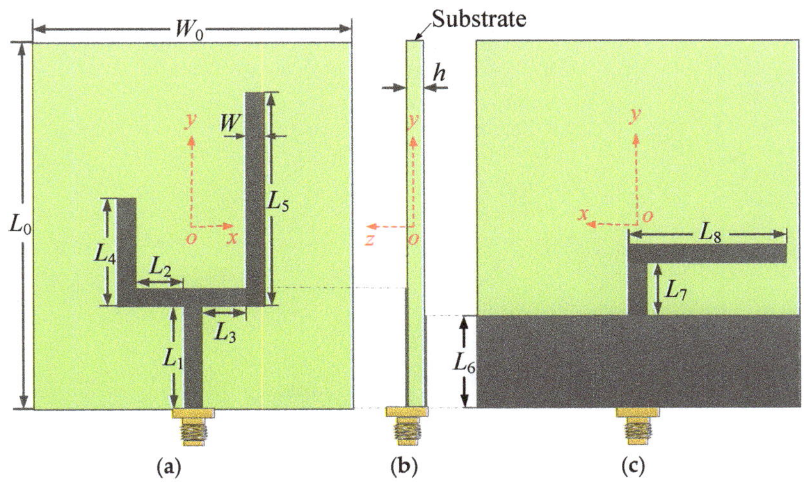

Figure 1. Geometrical design of the proposed antenna. (**a**) Top view. (**b**) Side view. (**c**) Bottom view.

Table 2. Dimensions of the proposed antenna (all values are in mm).

Parameter	Value	Parameter	Value	Parameter	Value
L_0	45	L_1	12	L_5	26
W_0	40	L_2	6.5	L_6	11
W	3	L_3	6.5	L_7	6
h	1.6	L_4	16	L_8	20.5

3. Theory and Design

Here, in this section, geometrical progress to achieve the proposed antenna design is outlined as shown in Figure 2. In the beginning, a simple monopole strip is printed on the upper surface of the FR-4 substrate whose one end is connected with SMA connector, as shown in Figure 2a. To achieve omnidirectional operating characteristics in the XOZ plane, a partial ground plane is taken on backside of the substrate (Ant. I).

For a microstrip monopole on a substrate, the wave is transmitted through both the medium and free space, thus the actual wavelength should be between the conduction wavelength of the medium and the operating wavelength of the free space. In this case, the electrical length L_{m} of the microstrip monopole can be calculated by [28,29]:

$$\frac{1}{2}\left(\frac{c}{4f_o\sqrt{\varepsilon_r}}+\frac{c}{4f_o\sqrt{\frac{\varepsilon_r+1}{2}+\frac{\varepsilon_r-1}{2}\left(1+\frac{12h}{W_m}\right)^{-0.5}}}\right)<L_m<\frac{c}{4f_o} \quad (1)$$

where c is the speed of light, f_o is the operating frequency, h is the thickness of the substrate, ε_r is the dielectric constant of the substrate, and W_m is the width of the monopole.

According to Equation (1), in order to cover the operating band of 2.2 GHz, the electrical length L_m of the microstrip monopole should be between 17.5 mm and 34.1 mm. At the same time, the antenna's higher-order operating mode should also cover 5.5 GHz frequency points. We optimized the parameters of the monopole in the CST simulator to meet both of the above requirements. The final values of L_m and W_m were 34 mm and 3 mm, respectively, and the dimension of the partial ground plane was 40×11 mm^2. Figure 3 shows the surface current distribution of the antenna under two different operating modes, including the surface current J_{sm1}, J_{sg1}, J_{sg2} in low-order operating mode and the surface current J_{sm2}, J_{sm3}, J_{sg3}, J_{sg4} in high-order operating mode. Due to the symmetrical structure of antenna I, J_{sg1} and J_{sg2} present mirror image distribution and so does J_{sg3} and J_{sg4}. It can be seen that the path length of J_{sm1} was about twice that of J_{sm2}, and the path length of

J_{sg1} was more than twice as much as that of J_{sg3}. Therefore, the frequency of the antenna in high-order mode should be more than twice that in low-order mode and higher than 4.4 GHz. The analysis results were consistent with the simulation results in Figure 4. We can see that the proposed monopole antenna (Ant I) operated at 2.22 GHz (2.1–2.34 GHz, S11 = −12.9 dB) and 5.51 GHz (5.34–5.73 GHz, S11 = −24.8 dB). This fabricated prototype showed good dual frequency performance and a simple structure, but its drawback was that the operating frequency did not fully cover the 5.25 GHz bands. If we optimize the antenna structure so that its high-frequency operating band covers 5.25 GHz, the volume of the antenna will increase, and we will need to change the antenna's prototype.

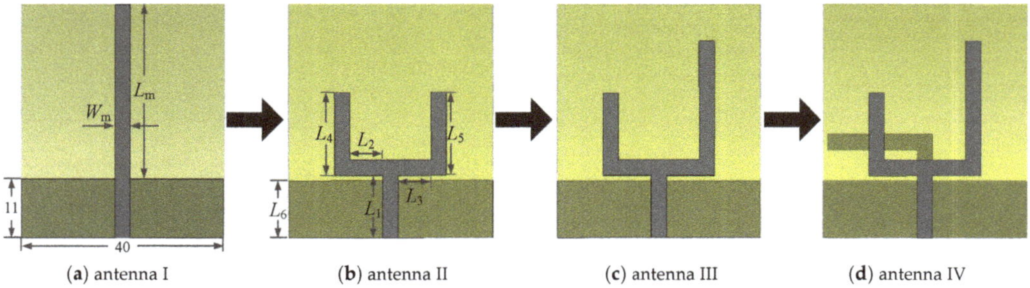

Figure 2. Evolution stages of the proposed antenna.

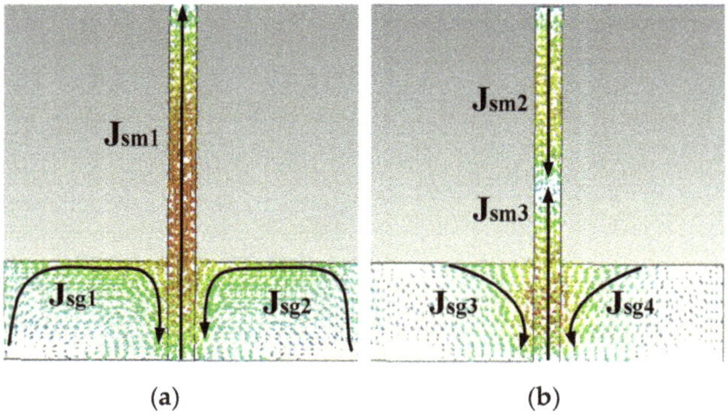

Figure 3. Surface current of the proposed monopole (Ant. I): (**a**) low-order mode; (**b**) high-order mode.

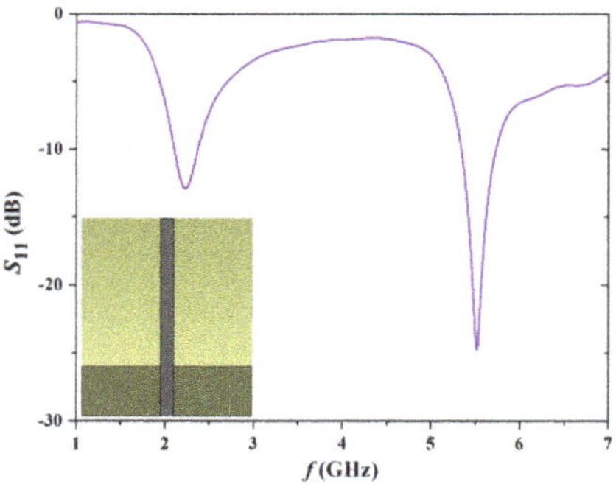

Figure 4. Simulated reflection coefficient of the proposed antenna I in Figure 2 (L_m = 34 mm, W_m = 3 mm).

Then, the monopole strip (Ant. I) is converted to a symmetrical turning fork-shaped patch antenna (STPA) with a rectangular ground, which is referred to Ant. II, shown in Figure 2b. One advantage of the dual-L-shaped design is that the L-shaped structure can increase the equivalent electrical length of the antenna in a finite volume, which will make the operating frequency of the antenna move to the low frequency band. Additionally, the dual-L structure is equivalent to a double element array, which can effectively enhance the radiation of the antenna and improve the S_{11} characteristics of the antenna. To make the comparison more meaningful and simplify the design process, we adopted the following principles: (1) the overall dimensions and ground size of the STPA shall be consistent with the design of Ant. I; (2) only the length of L_4 was changed to optimize the operating frequency band of STPA, while keeping other parameters constant. According to the above design principles, the parameters were designed as follows: L_1 = 12 mm, L_2 = L_3 = 6.5 mm, L_6 = 11 mm. Then, we optimized the operating frequency band of the antenna by adjusting the length of L_4 and L_5 in the U-shaped structure (L_4 = L_5).

Figure 5 shows the change in $|S_{11}|$ over the operating frequency range for step 2 with the change in L_4. It can be seen that the operating band always decreased significantly with the increasing L_4. When the length of L_4 was equal to 16 mm, this design had a good wideband characteristic with an impedance BW ($S_{11} < -10$ dB) covering 3.22–6.54 GHz, which covered the LTE (3.8-GHz), WLAN (5.2/5.8-GHz) and WiMAX (3.5/5.5-GHz) bands. Furthermore, the S_{11} at the resonant point of the antenna was almost better than −30 dB, which verified the radiation-enhanced effect of the double-L structure proposed above.

Figure 6 exhibits the simulated radiation patterns of the proposed STPA at two orthogonal cut planes. The calculated gains were 3.7, 5.5, and 5.9 dBi at 3.5, 5.2, and 5.8 GHz, respectively. It was concluded that the STPA antenna could meet most frequency bands of LTE, WLAN, and WiMAX applications based on the S_{11} characteristics and radiation direction characteristics. However, it's worth noting that this design misses the LTE (2.3-GHz), WLAN (2.4-GHz), and WiMAX (2.5-GHz) bands. It is possible to further increase the length of L_4 to cover 2.3~2.5 GHz, but this approach will cause the operating band to shift to lower frequencies and miss the high frequency point of the operating band, as the green line (L_4 = 20 mm) shows in Figure 5.

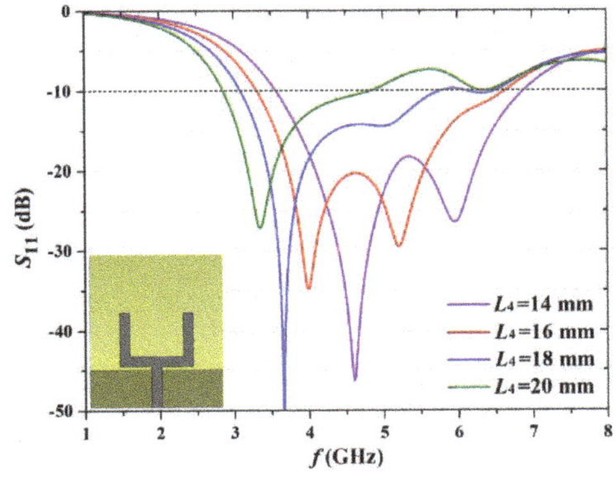

Figure 5. Simulated reflection coefficient of the STPA as a function of microstrip-line length L_4.

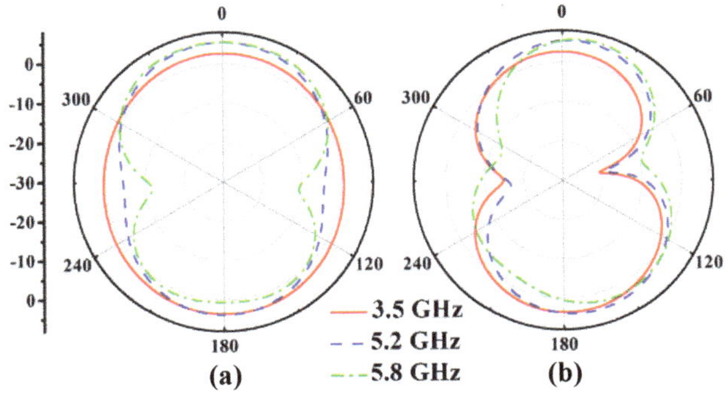

Figure 6. Simulated radiation patterns of the proposed STPA at (**a**) *xoz*-plane, (**b**) *yoz*-plane.

In order to improve its return loss of low frequency (2–3 GHz), we take the method of asymmetric designing and only increase the length of L_5, which can not only broaden the operating band by Superposition of resonant frequency points but also avoids the whole band moving to the low frequency region caused by increasing the length of both sides (L_4 and L_5). Thus, the symmetrical turning fork-shaped patch antenna (STPA) evolves into the asymmetrical turning fork-shaped patch antenna (ATPA).

Figure 7 shows the change in $|S_{11}|$ over the operating frequency range with the change in L_5 of the new design. It can be seen that the ATPA prototype shows the excellent characteristics of ultra-wideband. When the L_5 is optimized and properly selected as 26 mm, the ATPA achieves a −10 dB bandwidth of 5.05 GHz (2.95–7 GHz), which is enough for the majority of frequency points of wireless application. Figure 8 illustrates the calculated radiation patterns of the ATPA (L_5 = 26 mm) in the xoz and yoz planes under 2.4, 3.5, 5.2, and 5.8 GHz, and the corresponding simulated gains (L_5 = 26 mm) can reach 1.37, 3.62, 4.84, and 4.76 dBi, respectively. In the xoz plane, the radiation pattern is omnidirectional at 2.4, 3.5, and 5.2 GHz and shows a little bit of inomnidirectivity at 5.8 GHz. It can be seen that the radiation patterns are comparatively stable within the operating band and the same to radiation patterns in the yoz plane, which is convenient for the practical application.

The only drawback is that the S11 around 2.4 GHz only reaches −2.92 dB and still needs to be improved. What we should be concerned about is that continuing to increase the electrical length may enhance its low-frequency transmission characteristics, but it will result in a larger antenna size, which is clearly not an optimal approach. Thus, we need to find a new path to solve it.

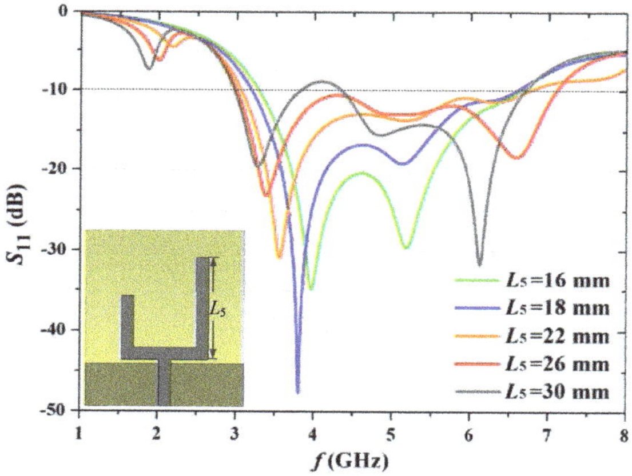

Figure 7. Simulated reflection coefficient of the ATPA as a function of microstrip-line length L_5. (L_1 = 12 mm, L_2 = L_3 = 6.5 mm, L_4 = 16 mm, L_6 = 11 mm, 45 × 40 mm^2).

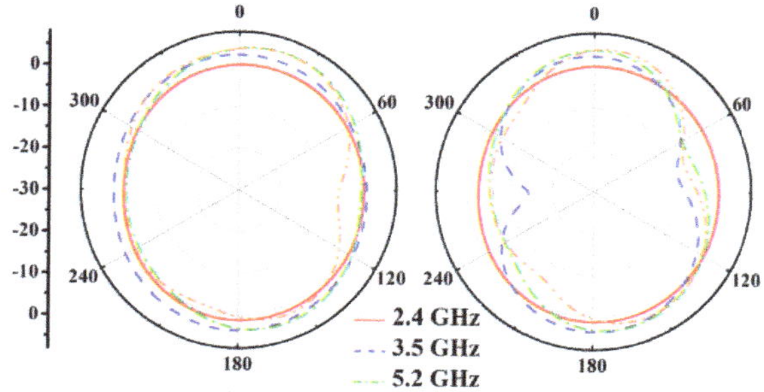

Figure 8. Simulated radiation patterns of the proposed ATPA at (**a**) *xoz*-plane, (**b**) *yoz*-plane.

For the ATPA, there are two main current paths on either side of the asymmetric tuning fork structure, the equivalent electrical lengths of which are approximately equal to (L2 + L4) and (L3 + L5), respectively. According to the formula [17]: f = c/4 L (c is the speed of light in free space; and L is the electrical length.), it can be calculated that the two resonant points of the antenna are about 2.31 and 3.3 GHz, which are consistent with the simulation results in Figure 7. Therefore, we need to design a new current path with a length between 22.5 mm and 32.5 mm to enhance the low frequency resonance characteristics. However, the return loss only reaches −10 dB in the range of 5 to 6 GHz, and thus the introduction of a new structure is likely to worsen the return loss in this range. Thus, we need to subtly design a new structure, which could not only improve the reflection characteristics of both low and high frequency regions but also maintain the current size.

Based on the analysis above, we proposed a novel tuning fork-shaped tri-band planar antenna (NTTPA) by meticulously adding an inverted L-shaped radiation patch on the ground of the ATPA. In addition, the L-shaped patch is designed on the short side of the turning fork-shaped patch to compensate for the radiation asymmetry caused by the asymmetric turning fork structure. The antenna is optimized by adjusting L_7 and L_8, and the final structure size is shown in Table 1.

The simulated surface current distributions under 2.4 and 5.8 GHz are given in Figure 9 to illustrate the operating mechanism for the ATPA and NTTPA. The larger value of the current distribution is indicated in red, and the smaller value is in blue. It can be seen that the introduction of the inverted L-shaped structure not only provides a new current path but also enhances the circuit density in the existing paths on the asymmetric turning fork patch. Additionally, the same effect could be observed at 5.8 GHz.

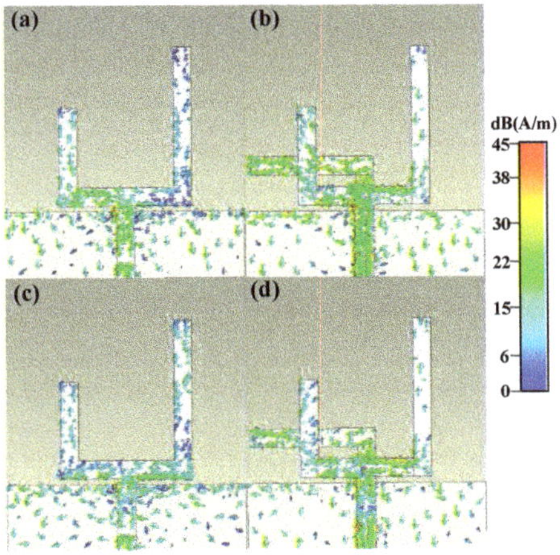

Figure 9. Simulation results of the surface current distributions. (**a**) ATPA at 2.4 GHz. (**b**) NTTPA at 2.4 GHz. (**c**) ATPA at 5.8 GHz. (**d**) NTTPA at 5.8 GHz.

Figure 10 illustrates the S11 characteristic of different designs from the microstrip monopole to the NTTPA (Antenna I refers to the monopole, Antenna II refers to the STPA, Antenna III refers to the ATPA, and Antenna IV refers to the NTTPA). It is obvious that the addition of inverted L-shaped structures reduces the low frequency return loss, which is consistent with the above analysis above. The simulation results shows that −10 dB bandwidth of the proposed novel tuning fork-shaped tri-band patch antenna (NTTPA) covers 2.28–3.7 and 4.7–6.2 GHz, which is enough for LTE (2.3–2.4, 3.6–3.8), WLAN (2.4–2.484, 5.15–5.35, 5.725–5.825), and WiMAX (2.5–2.69, 3.4–3.69, 5.25–5.85) applications.

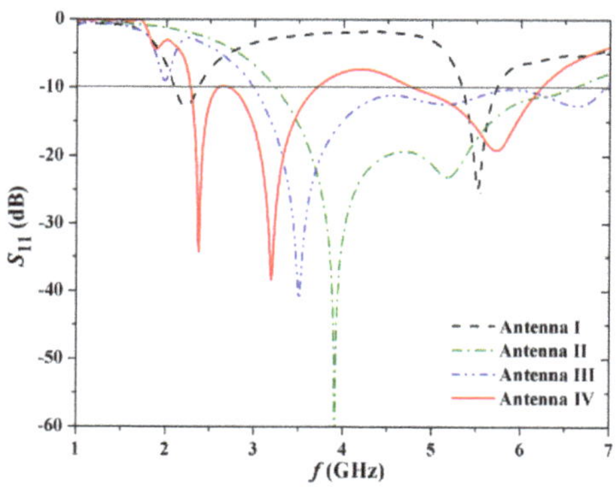

Figure 10. Simulated reflection coefficient of different proposed antenna.

4. Results and Discussion

To confirm the predicted radiation-enhanced and multiband operating characteristics exhibited in Figures 9 and 10, a novel tuning fork-shaped tri-band planar antenna (NTTPA) was fabricated and tested. The physical picture of the proposed antenna is shown in Figure 11. The performance of the proposed antenna was tested in an anechoic chamber, shown in Figure 12, including return loss, gain, and radiation pattern. The measured return loss tested by Agilent vector network analyzer-E8363B was compared by the simulation results. As shown in Figure 13, a strong agreement was observed. The measured impedance BW over −10 dB was 17.8% (2.2–2.63 GHz), 32.8% (2.73–3.8 GHz), and 20.5% (5.13–6.3 GHz), and the maximum value could reach −20.9 dB (at 2.42 GHz), −24.5 dB (at 3.29 GHz), and −23.1 dB (at 5.98 GHz). It can be noted that the −9.8 dB impedance BW covered 2.2–3.8 GHz with a percentage BW of 53.3%.

Figure 14 shows the calculated and measured gains for the proposed antenna. The measured ranges of the antenna gains were 0.8–1.9, 2–3.6, and 4.7–5.6 dBi in Band-I (2.2–2.63 GHz), band-II (2.73–3.8 GHz), and band-III (5.13–6.3 GHz), respectively. Small variations were observed between calculated and test results for both bands, which can be attributed to possible cable effect and tolerance in measurement.

Figure 15 shows the normalized radiation patterns at 2.4, 3.5, 5.2, and 5.8 GHz, respectively. The difference between the simulated and experimental results was small. We can see that the proposed NTTPA showed good omnidirectional radiation property in the xoz plane and directional radiation property in the yoz plane. It is worth noting that the radiation direction of the antenna was very close at a different frequency, which is very conducive to the industrial application. The efficiency and front-to-back ratio plots are displayed in Figures 16 and 17, respectively. Compared with the same tri-band antenna in the references, the S11 proposed in this paper is lower [19,20].

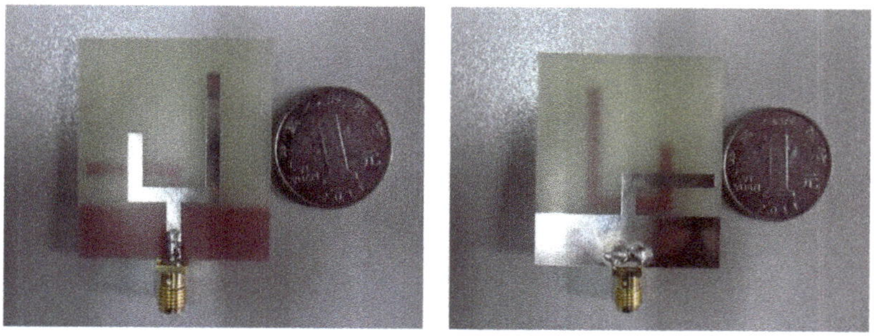

Figure 11. The physical picture of the proposed antenna.

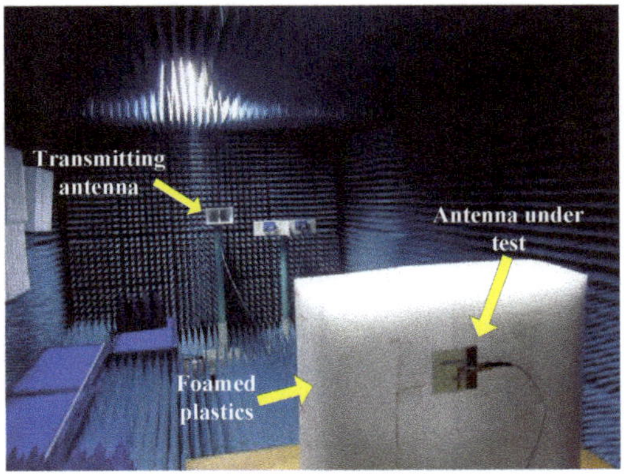

Figure 12. Calculated and experimental S_{11} of the proposed NTTPA.

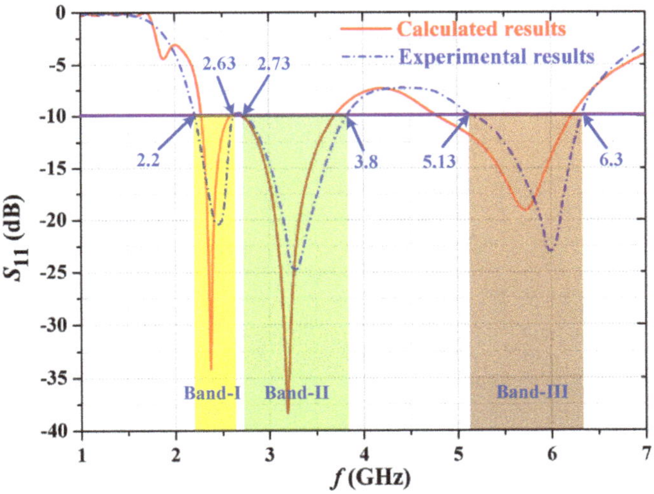

Figure 13. Calculated and experimental S11 of the proposed NTTPA.

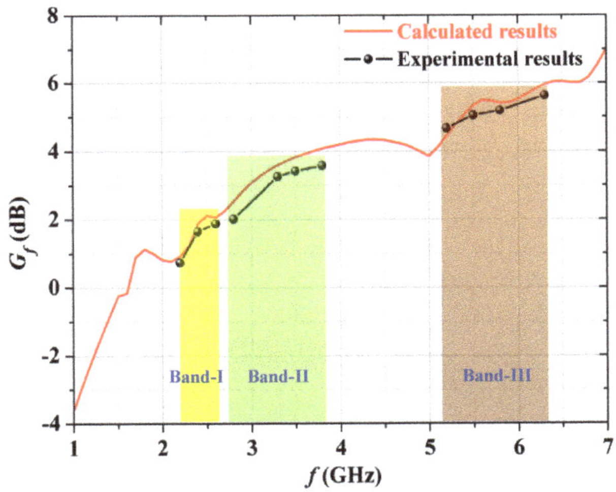

Figure 14. Calculated and experimental Gain of the proposed NTTPA.

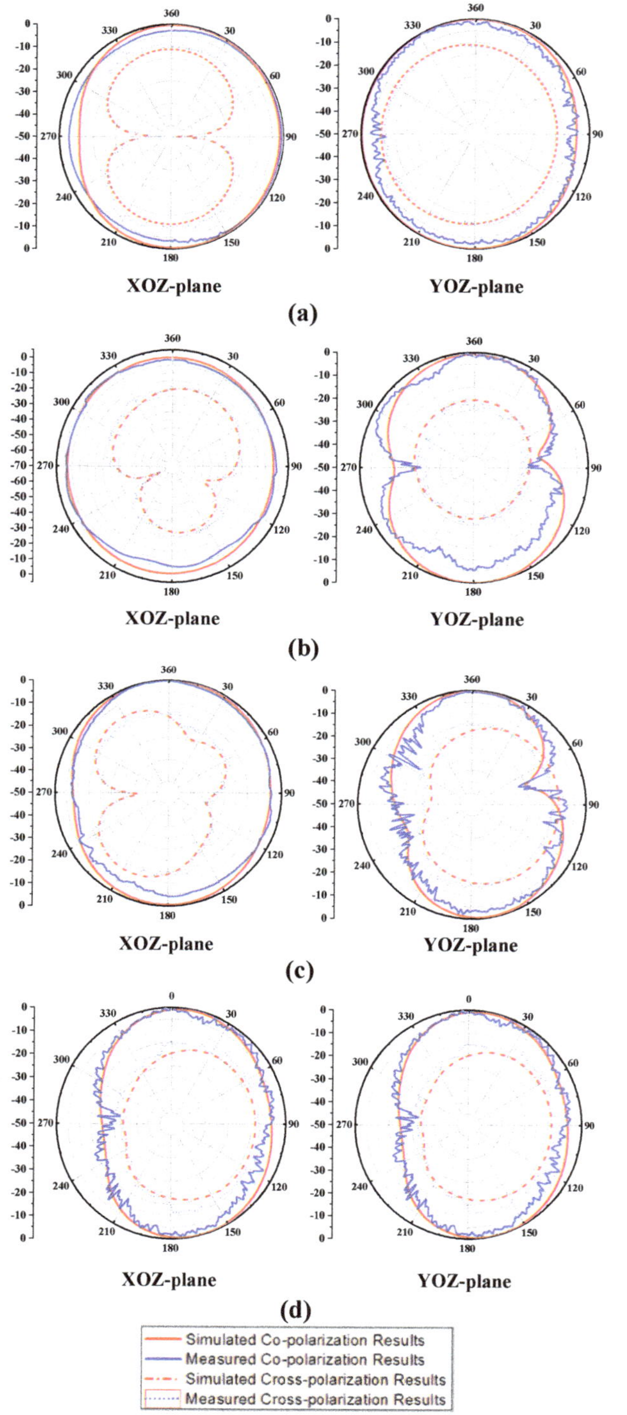

Figure 15. Simulated and measured radiation patterns in x-o-z-plane and y-o-z-plane at: (**a**) 2.4 GHz, (**b**) 3.5 GHz, (**c**) 5.2 GHz, (**d**) 5.8 GHz.

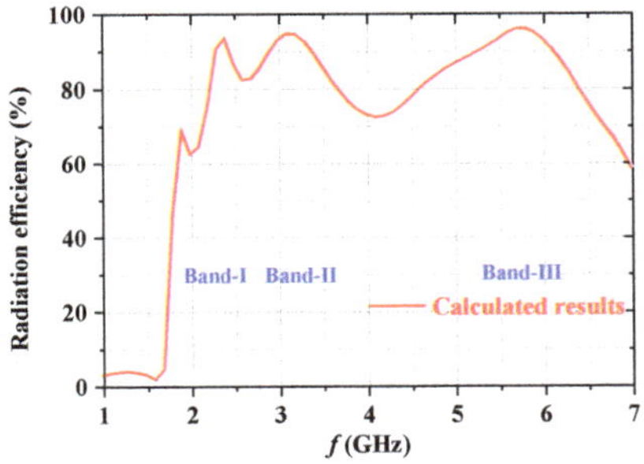

Figure 16. Radiation efficiency of the proposed antenna.

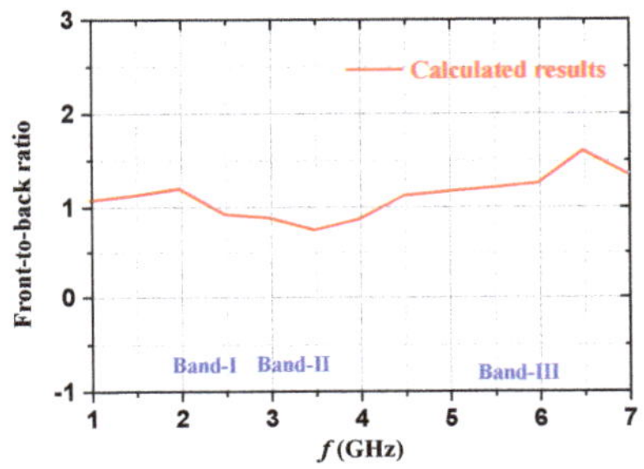

Figure 17. Front-to-back ratio of the proposed antenna.

5. Conclusions

A novel tuning fork-shaped tri-band planar antenna (NTTPA) for LTE (2.3/3.8-GHz), WLAN (2.4/5.2/5.8-GHz), and WiMAX (2.5/3.5/5.5-GHz) applications has been successfully reported with a compact size of 45 mm × 40 mm. The evolution of achieving this proposed antenna has been discussed in detail. The antenna mentioned in this paper can keep good performance in three frequency bands while realizing the miniaturization of its size. It has a wide bandwidth and a high gain, relatively. Besides exhibiting desirable impedance BW and overall size, the proposed antenna has also shown very stable radiation patterns across the three bands of interest, which is suitable for wireless communication applications. At the same time, it also provides useful inspiration for 5G antenna designs. Work can be performed in the future to further realize miniaturization.

Author Contributions: Data curation, J.D. and J.S.; Funding acquisition, W.C.; Investigation, Q.L.; Methodology, Q.L. and C.G.; Resources, J.F.; Supervision, J.D.; Visualization, W.C. and T.L.; Writing original draft, Q.L.; Writing review and editing, J.F., J.D. and T.L. All authors have read and agreed to the published version of the manuscript.

Funding: This work was supported by the Foundation Enhancement Plan (2017-JCJQ-ZD-02), National Key Laboratory of Science and Technology on Space Microwave (No: 6142411332211), the National Natural Science Foundation of China (61803042), and the Fundamental Research Funds for the Central Universities, CHD (300102322103).

Data Availability Statement: Not applicable.

Conflicts of Interest: The authors declare no conflict of interest.

References

1. Carver, K.; Mink, J. Microstrip antenna technology. *IEEE Trans. Antennas Propag.* **1981**, *29*, 2–24. [CrossRef]
2. Dey, S.; Mittra, R. Compact microstrip patch antenna. *Microw. Opt. Technol. Lett.* **1996**, *13*, 12–14. [CrossRef]
3. Li, J.; Guo, J.; He, B.; Zhang, A.; Liu, Q.H. Tri-band CPW-fed stub-loaded slot antenna design for WLAN/WiMAX applications. *Frequenz* **2016**, *70*, 521–526. [CrossRef]

4. Yildirim, B.; Basaran, E.; Turetken, B. Dielectric-Loaded Compact WLAN/WCDMA Antenna With Shorted Loop and Monopole Elements. *IEEE Antennas Wirel. Propag. Lett.* **2013**, *12*, 288–291. [CrossRef]
5. Bakariya, P.S.; Dwari, S.; Sarkar, M.; Mandal, M.K. Mandal. Proximity-coupled multiband microstrip antenna for wireless applications. *IEEE Antennas Wirel. Propag. Lett.* **2015**, *14*, 646–649. [CrossRef]
6. Malik, J.; Patnaik, A.; Kartikeyan, M.V. A Compact Dual-Band Antenna With Omnidirectional Radiation Pattern. *IEEE Antennas Wirel. Propag. Lett.* **2015**, *14*, 503–506. [CrossRef]
7. Li, W.-W.; Li, Q.-H.; Meng, Y.; Wang, J.-Y.; Xu, W.-M. A broadband microstrip patch antenna with multiple open slots. *Microw. Opt. Technol. Lett.* **2019**, *61*, 626–632. [CrossRef]
8. Sim, C.Y.D.; Chen, H.D.; Yeh, C.H.; Lin, H.L. Small size triple band monopole antenna with a parasitic element. *Microw. Opt. Technol. Lett.* **2015**, *57*, 342–348. [CrossRef]
9. Li, L.; Zhang, X.; Yin, X.; Zhou, L. Compact Triple-Band Printed Monopole Antenna for WLAN/WiMAX Applications. *IEEE Antennas Wirel. Propag. Lett.* **2016**, *15*, 1853–1855. [CrossRef]
10. van Rooyen, M.; Odendaal, J.W.; Joubert, J. High-Gain Directional Antenna for WLAN and WiMAX Applications. *IEEE Antennas Wirel. Propag. Lett.* **2017**, *16*, 286–289. [CrossRef]
11. Liu, G.; Liu, Y.; Gong, S. Compact tri-band wide-slot monopole antenna with dual-ring resonator for WLAN/WiMAX applications. *Microw. Opt. Technol. Lett.* **2016**, *58*, 1097–1101. [CrossRef]
12. Yang, X.; Ji, Y.; Ge, L.; Zeng, X.; Wu, Y.; Liu, Y. A dual-band radiation-differentiated patch antenna for future wireless scenes. *IEEE Antennas Wirel. Propag. Lett.* **2020**, *19*, 1007–1011. [CrossRef]
13. Feng, S.; Zhang, L.; Yu, H.-W.; Zhang, Y.-X.; Jiao, Y.-C. A Single-Layer Wideband Differential-Fed Microstrip Patch Antenna With Complementary Split-Ring Resonators Loaded. *IEEE Access* **2019**, *7*, 132041–132048. [CrossRef]
14. Li, W.; Xia, Z.; You, B.; Liu, Y.; Liu, Q.H. Dual-Polarized H-Shaped Printed Slot Antenna. *IEEE Antennas Wirel. Propag. Lett.* **2017**, *16*, 1484–1487. [CrossRef]
15. Yang, X.; Kong, F.; Liu, X.; Song, C. A CPW-fed triple-band antenna for WLAN and WiMAX applications. *Radioengineering* **2014**, *23*, 1086–1091.
16. Chaturvedi, D.; Kumar, A.; Raghavan, S. An integrated SIW cavity-backed slot antenna-triplexer. *IEEE Antennas Wirel. Propag. Lett.* **2018**, *17*, 1557–1560. [CrossRef]
17. Mao, X.; Zhu, Y.; Li, Y. A Simple Tri-Band Proximity Coupling Fed Compact Antenna for 2.7GHz, WLAN and Sub-6GHz Communication Applications. In Proceedings of the 2018 IEEE International Symposium on Antennas and Propagation & USNC/URSI National Radio Science Meeting, Boston, MA, USA, 8–13 July 2018; pp. 1013–1014.
18. Boutejdar, A.; Halim, B.I. Design of a Compact Tri-band Ring Antenna Using Two Parasitic Ring Resonators and Partial Ground Plane for WiMAX and RADAR Applications. In Proceedings of the 2019 IEEE International Electromagnetics and Antenna Conference (IEMANTENNA), Vancouver, BC, Canada, 17–19 October 2019; pp. 51–55.
19. Kiani, S.H.; Khan, M.A.; Rafique, U.; Marey, M.; Alharbi, A.G.; Mostafa, H.; Khan, M.A.; Abbas, S.M. High Performance Eight-Port Dual-Band MIMO Antenna System for 5G Devices. *Micromachines* **2022**, *13*, 959. [CrossRef] [PubMed]
20. Kiani, S.H.; Marey, M.; Savci, H.; Mostafa, H.; Rafique, U.; Khan, M.A. Dual-Band Multiple-Element MIMO Antenna System for Next-Generation Smartphones. *Appl. Sci.* **2022**, *12*, 9694. [CrossRef]
21. Zahid, M.N.; Gaofeng, Z.; Kiani, S.H.; Rafique, U.; Abbas, S.M. H-Shaped Eight-Element Dual-Band MIMO Antenna for Sub-6 GHz 5G Smartphone Applications. *IEEE Access* **2022**, *10*, 85619–85629. [CrossRef]
22. Malallah, R.; Shaaban, R.M.; Al-Tumah, W.A.G. A dual band star-shaped fractal slot antenna: Design and measurement. *AEU-Int. J. Electron. Commun.* **2020**, *127*, 153473. [CrossRef]
23. Haque, S.M.; Alam, H. Miniaturized dual-band slot antenna design for GPS, amateur radio and WLAN applications. *Int. J. RF Microw. Comput.-Aided Eng.* **2020**, *30*, e22125. [CrossRef]
24. Madhav, B.T.P.; Monika, M.; Kumar, B.M.S.; Prudhvinadh, B. Dual band reconfigurable compact circular slot antenna for WiMAX and X-band applications. *Radioelectron. Commun. Syst.* **2019**, *62*, 474–485. [CrossRef]
25. Madhavi, D.; Jagadeesh, D. Dual-band semi-hexagonal slot antenna backed by SIW for WLAN/WBAN applications. *Prog. Electromagn. Res. C* **2022**, *121*, 221–232. [CrossRef]
26. Kumar, A.; Jhanwar, D.; Sharma, M.M. A compact printed multistubs loaded resonator rectangular monopole antenna design for for multiband wireless systems. *Int. J. RF Comput. Aided Eng.* **2017**, *27*, e21147. [CrossRef]
27. Kumar, A.; Jhanwar, D.; Sharma, M.M. Miniaturized multistubs loaded rectangular loaded monopole antenna for multiband applications based on theory of characteristics modes. *Prog. Electromag. Res. C* **2019**, *92*, 177–189. [CrossRef]
28. Balanis, C.A. *Antenna Theory: Analysis and Design*; Wiley: New York, NY, USA, 2005.
29. Ghouz, H.H.M.; Sree, M.F.A.; Ibrahim, M.A. Novel Wideband Microstrip Monopole Antenna Designs for WiFi/LTE/WiMax Devices. *IEEE Access* **2020**, *8*, 9532–9539. [CrossRef]

electronics

MDPI

Communication

Pre-Design of Multi-Band Planar Antennas by Artificial Neural Networks

Mohamed Aziz Lahiani [1], Zbyněk Raida [2,3], Jiří Veselý [3] and Jana Olivová [3,*]

[1] National Institute of Applied Sciences and Technology, Tunis 625 00, Tunisia
[2] Faculty of Electrical Engineering and Communication, Brno University of Technology, Brno-Královo Pole, 616 00 Brno, Czech Republic
[3] Faculty of Military Technology, University of Defence, Brno-Střed, 662 10 Brno, Czech Republic
* Correspondence: jana.olivova2@unob.cz

Abstract: In this communication, artificial neural networks are used to estimate the initial structure of a multiband planar antenna. The neural networks are trained on a set of selected normalized multiband antennas characterized by time-efficient modal analysis with limited accuracy. Using the Deep Learning Toolbox in Matlab, several types of neural networks have been created and trained on the sample planar multiband antennas. In the neural network learning process, suitable network types were selected for the design of these antennas. The trained networks, depending on the desired operating bands, will select the appropriate antenna geometry. This is further optimized using Newton's method in HFSS. The use of the neural pre-design concept speeds up and simplifies the design of multiband planar antennas. The findings presented in this paper will be used to refine and accelerate the design of planar multiband antennas.

Keywords: multi-band antennas; feed-forward neural network; cascade-forward neural network; probabilistic neural network; full-wave analysis

Citation: Lahiani, M.A.; Raida, Z.; Veselý, J.; Olivová, J. Pre-Design of Multi-Band Planar Antennas by Artificial Neural Networks. *Electronics* **2023**, *12*, 1345. https://doi.org/10.3390/electronics12061345

Academic Editor: Antonio Dourado

Received: 13 February 2023
Revised: 3 March 2023
Accepted: 10 March 2023
Published: 12 March 2023

1. Introduction

When designing a conventional multi-band planar antenna, a proper patch geometry has to be selected to obtain resonant frequencies in the requested operational bands [1]. In the next step, a full-wave numerical model of the antenna is developed in an electromagnetic simulator, and the model is optimized to meet the required parameters of the antenna as accurately as possible.

A proper antenna geometry is selected by a designer experienced with it. In this communication, we train artificial neural networks (ANN) to represent the designer's experience. ANNs are trained on a set of normalized multi-band antennas characterized by modal analysis, which efficiently produces approximate training patterns. The modal analysis itself and the way of creating training sets are described in Section 2.1.

When designing a planar antenna on a prescribed substrate for requested operation bands, normalization related to the wavelength in the substrate is performed, and normalized resonant frequencies are mapped by ANN to the optimum antenna geometry. For mapping, three types of ANN are used:

- Feed-forward back-propagation ANN [2].
- Cascade-forward back-propagation ANN [3].
- Probabilistic ANN [4].

In Section 2.2, ANN are briefly introduced, and the process of their training is described.

In Section 3, we present a design example to illustrate the functionality of the neural pre-design. Section 4 concludes the paper.

In the open literature, several papers on an ANN-enhanced antenna design have been published. The papers cover the following topics:

- On-line neural synthesis of radiation patterns. In an article about the design of a cognitive antenna array [5], the radiation pattern of a conformal patch array has been adapted to a complex environment by a proper phasing of elements. A used deep reinforcement learning was based on an on-line network, which updated parameters for training, and a target network, which calculated the loss function exploiting data from an experience pool. In a paper investigating the synthesis of conformal phased array antenna (PAA) patterns using deep synthesis [6], the on-line ANN and the target ANN created a tandem network structure that minimized the difference between the requested pattern and the current one.

In the described approaches, antenna geometries are fixed, and input signals are optimized to reach requested radiation patterns. In this communication, the neural network selects the optimum shape of the patch to form the optimum current distribution related to multiple resonances in the requested operational bands.

- Black-box modeling of antenna structures. Computer processing unit (CPU)-time moderate ANNs are trained to approximate the results of CPU-time expansive full-wave analysis over a limited definition space. This approach can be applied both to canonical structures [7] and advanced ones. A patch antenna with a ground plane defected by split-ring resonators was modeled by a multi-layer perceptron and optimized by a particle swarm algorithm in [8]. A high-gain quasi-Yagi antenna with a parabolic reflector was modeled by a pyramidal deep regression network in [9].

For multi-physics modeling of microwave filters, the authors used a deep hybrid neural network [10]. The network was conceived as a cascade of ANNs with different parameters. For resonant structures like filters, this modeling approach was shown to be beneficial.

Neural black-box models usually face the problem of sufficient generality and acceptable accuracy being reached with reasonable efforts [11]. A sufficiently general model should cover a sufficiently large definition space of an optimum dimension, which is given by the number of state variables. A sufficiently accurate model usually requires time-expansive simulations of training patterns.

In order to overcome these difficulties, we propose an approximate classifier that maps normalized operation frequencies to the optimum layout of an antenna element. The definition space is reduced by working with normalized frequencies and dimensions, and the CPU time needed to create training sets is minimized by using a time-moderate modal analysis. If the pre-designed structure is sufficiently accurate, then conventional local optimization can be quick and inexpensive.

- Antenna design by ANN and optimizer. If the efficient and accurate black-box model is completed by an optimization algorithm, a simple design tool can be developed. In [12], a patch is divided into pixels. The shape of the patch is synthesized by combining a convolutional ANN in the role of a forward model (geometry at the input and performance at the output) and a genetic algorithm in the role of the optimizer.

The described approach can be used even for a multi-objective design of antennas. In [13], a deep neural model was combined with Thomson sampling for efficient multi-objective optimization to reveal the Pareto front of optimal solutions.

In [14], planar ultra-wideband antennas were designed by combining deep structures (the cascade of an extreme learning machine, a deep belief network, and a restricted Boltzmann machine) and a particle swarm optimization as a global optimizer.

Since properly trained ANNs can provide the response very quickly (only a few arithmetic operations are needed to be executed in parallel), neural models are advantageous when combined with evolutionary algorithms [15,16], swarm-intelligence approaches [17,18], and other CPU-time expensive global optimizers. But the question of the development of a sufficiently general and sufficiently accurate neural model remains.

The approach presented in this communication is based on neural classifiers. A pure classification of planar microwave filters was presented in [19]. At the input of a deep network, a bitmap with a photo of the filter was introduced. ANN classified the filter as a low-pass or band-pass one and determined the order of the filter.

In [20], the classifier considered the requested gain, impedance bandwidth, and operation frequency and selected among a patch antenna, a spiral antenna, or a horn antenna. Then, the selected antenna was designed by an inverse neural model that mapped requested antenna parameters to antenna dimensions. Antenna geometries were given, and dimensions were computed.

In the presented approach, the operational frequencies of a planar multiband antenna are mapped by the trained neural network to the optimum normalized geometry of the patch. Consecutively, the selected geometry is modeled in the High Frequency Structure Simulator (HFSS) and optimized by the Newton algorithm. Thanks to the successful pre-design, the local optimization is sufficient, and a low number of iteration steps is required.

This provides further acceleration of the antenna structure design. According to our knowledge, the described exploitation of ANN for the antenna pre-design has not been presented in the open literature yet.

2. Methods

We intend to create an efficient tool for the design of planar multi-band antennas. In order to ensure a sufficient design efficiency, we use a classifier for the selection of an optimal normalized antenna geometry, which is the first step. Second, the geometry is denormalized, taking into account the wavelength of the used substrate, and a full-wave HFSS model is developed. Finally, the model is optimized by the local Newton optimizer in a few steps.

In the following paragraphs, the individual design steps are described in detail.

2.1. Training Sets by Modal Analysis

The normalized antenna geometries consider an air substrate with the dielectric constant $\varepsilon_r = 1$, negligible loss $\tan \delta = 0$, and a thickness of $h = 1$ mm. Since the corresponding wavelength is $\lambda_0 = 300$ mm, we create a set of slotted patches with the fundamental dimension 150 mm × 150 mm.

Geometries of training patterns are depicted in Figure 1, with pixel dimensions of 5 mm × 5 mm. The green pixels represent metallization (a perfect electric conductor in an approximation, a standard copper foil in an implementation), and the white pixels represent an uncovered substrate.

The size of the patch is fixed at 30 × 30 pixels except for the extended patch antenna (Figure 1b). The width of slots is fixed at 1 pixel, except for the square slot (Figure 1i).

Training geometries are going to cover the most important mechanisms of exciting multi-band behavior:

- The conventional patch (Figure 1a) plays the role of reference. The multi-band behavior corresponds to the higher modes of the patch. The extended patch antenna (Figure 1b) can be understood as a conventional patch with a capacitive prolongation [21].
- The L-slot antenna (Figure 1c) represents patches with a two-segment slot that breaks the patch edge into two parts. The currents in both sub-areas of the patch are galvanically connected [21].
- The U-slot antenna (Figure 1d) represents patches with a three-segment slot that breaks the patch edge into two parts. The currents in both sub-areas of the patch are separated [21].
- The T-slot antenna (Figure 1e) can be understood as a folded slot dipole fed by a coplanar waveguide (CPW) [22]. A magnetic current flowing in the slot loop is the source of the radiation.

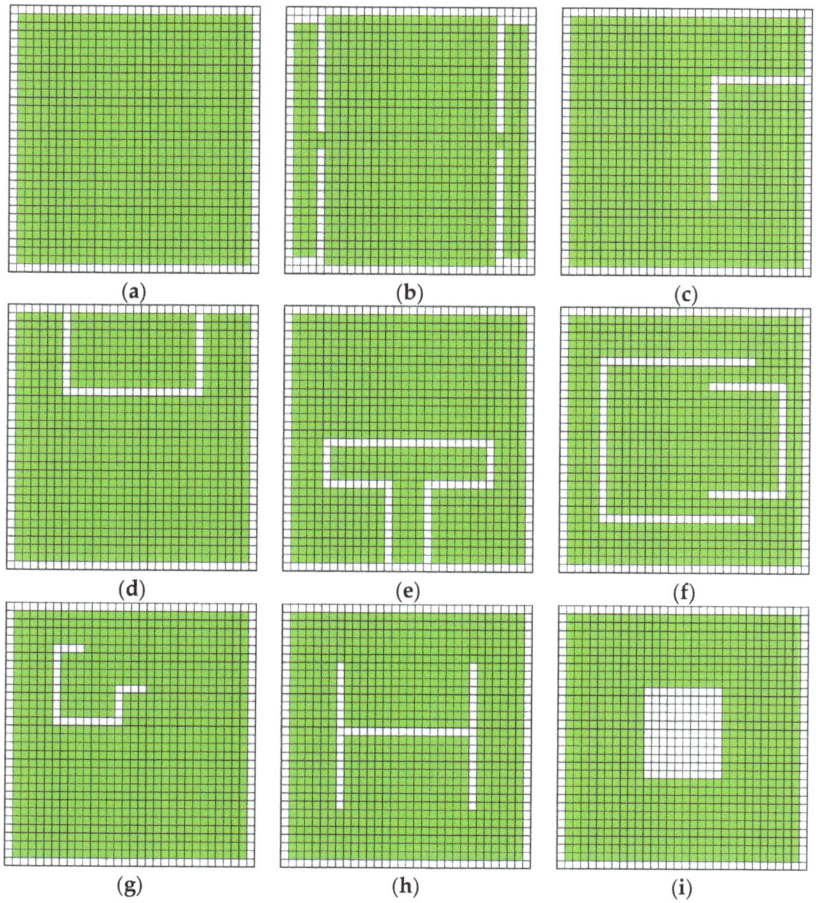

Figure 1. Training geometries of normalized multi-band patches: (**a**) conventional patch, (**b**) extended patch, (**c**) L-slot patch, (**d**) U-slot patch, (**e**) T-slot patch, (**f**) double U-slot patch, (**g**) G-slot patch, (**h**) H-slot patch, and (**i**) square-slot patch. Pixel dimensions: 5 mm × 5 mm. Green: conductive surface; white: uncovered substrate.

The double-U antenna, the G antenna, and the H antenna are patches with slots inside the antenna element. These slots directly influence the current distribution on the patch and the multiband behavior of the patch. The double-U antenna (Figure 1f) consists of a large radiator (the whole patch) and a small one (the area inside slots), which define two operational bands [21]. The G antenna (Figure 1g) divides the patch into an internal area and two external ones, potentially offering a triple-band operation [23]. In the case of a non-symmetrically located non-symmetrical H slot (Figure 1h), more than three bands can be created [24]. Finally, a square slot (Figure 1i) creates the equivalent of a loop antenna [25].

With varying positions and dimensions of slots, 60 training layouts were created. Each layout was modeled in the partial differential equation (PDE) tool of MATLAB [26]:

- Metallic parts of the layout were enclosed by Neumann boundary conditions.
- The solver was set to evaluate eigenmodes.
- Eigenvalues were considered within the interval $<0; 5 \times 10^4>$.

Since eigenvalues are equal to the squared wave number, the operational frequencies of antennas can be evaluated according to [27]:

$$f_n = \frac{c}{2\pi}\sqrt{a_n} \tag{1}$$

where f_n is the n-th resonant frequency of the antenna, corresponding to the n-th eigenvalue a_n computed by the PDE tool, and c is the velocity of light in vacuum.

The numerical model of the G-slot patch is depicted in Figure 2. Boundaries (black lines) are associated with the Neumann boundary condition. The dimensions of the patch are fixed. Position and dimensions of the slot are varied by changing the coordinates of polygon vertices A through L.

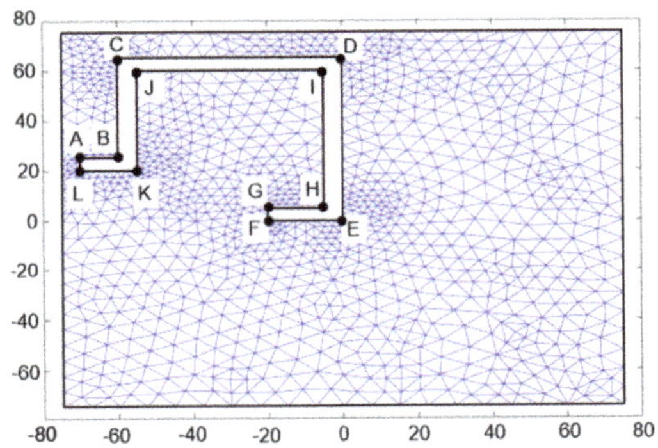

Figure 2. Modal analysis of the G-slot patch antenna by the finite-element method in the PDE tool of MATLAB.

The three lowest modes computed by the PDE tool (Figure 3) were used to compose a training pattern. Other training patterns were prepared in a similar way.

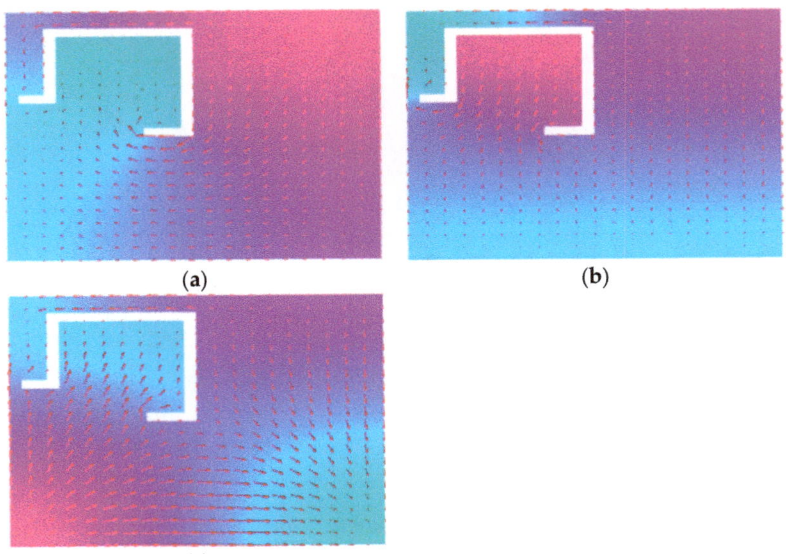

Figure 3. The lowest 3 modes of the G-slot antenna were computed by the finite-element method in the PDE tool in MATLAB.

As a result, we have 60 triplets of resonant frequencies for 60 patch layouts. Neural networks are trained to map triples of frequencies on the input to the index of a corresponding patch layout on the output. Details are given in the next paragraph.

2.2. ANN and Training Process

We use neural networks to map triplets of resonant frequencies computed by the modal analysis to the index of a corresponding patch layout. Hence, the networks have three neurons in the input layer and a single neuron in the output layer. The input neurons simply distribute signals to neurons in the hidden layer. The output neuron collects signals from neurons in the hidden layer and processes the signal using the activation function [1].

In order to select optimal patch layouts, we used 3 ANNs from the Deep Learning Toolbox of MATLAB [28]:

- Feed-forward back-propagation network. Input patterns are sequentially introduced to input neurons, the ANN response is computed, and the difference (an error) between the output and the response from the training set is evaluated. The error propagates back to the input and changes the settings of neurons to minimize the error.

When implementing the network, we used TRAINLM as a training function, LEARNGND as a learning function, MSE as a performance function, and TANSIG as an activation function. Training details are given in Figure 4a.

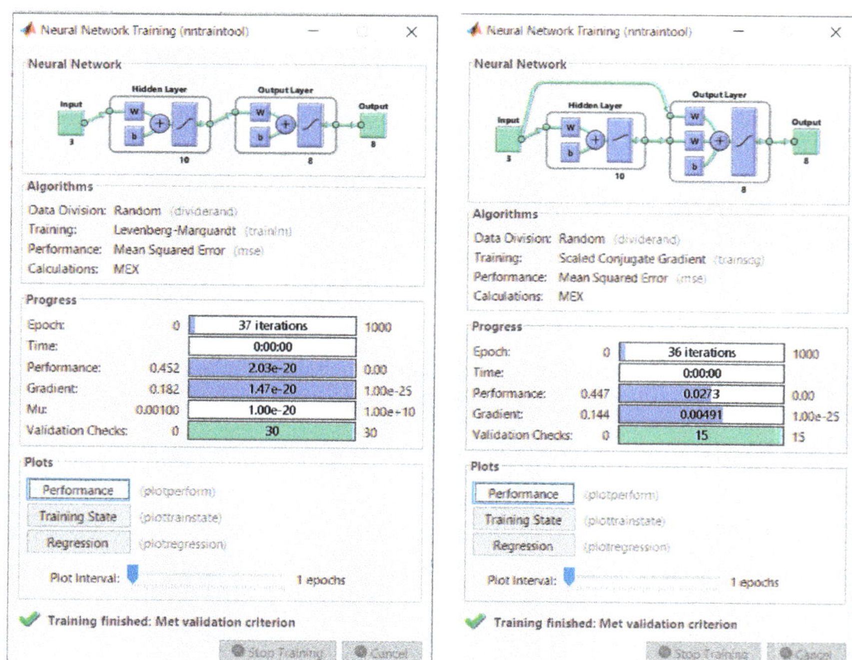

(a) (b)

Figure 4. Training neural classifiers in the Deep Learning Toolbox of MATLAB: (**a**) feed-forward back-propagation ANN; (**b**) cascaded-forward back-propagation ANN.

- A cascade-forward back-propagation network is similar to a feed-forward network, but includes connections from the input and every previous layer to the following layers. The network accommodates the nonlinear relationship between the input and the output but does not eliminate the linear relationship in between.

When implementing the network, we used TRAINSCG as a training function, LEARNGND as a learning function, MSE as a performance function, and LOGSIG as an activation function. Training details are given in Figure 4b.

- A probabilistic network contains radial neurons with a Gaussian activation function in the hidden layer. The output layer sums contributions for each class of input patterns, producing a vector of probabilities as the output. The transfer function of the output layer picks the maximum of these probabilities and produces 1 for the corresponding class. For other classes, 0 is produced.

When implementing the network, we used NEWPNN to create and train the ANN.

In order to test the quality of training, four antennas differing from training patterns (Figure 5) were created and analyzed. Corresponding triplets of resonant frequencies were introduced to the inputs, and ANN was asked to classify the optimum layout:

- Feed-forward ANN succeeded with 61.2%/59.3%/41.2%/61.4%;
- Cascaded-forward ANN succeeded with 63.2%/52.3%/40.0%/46.0%;
- The probabilistic ANN failed.

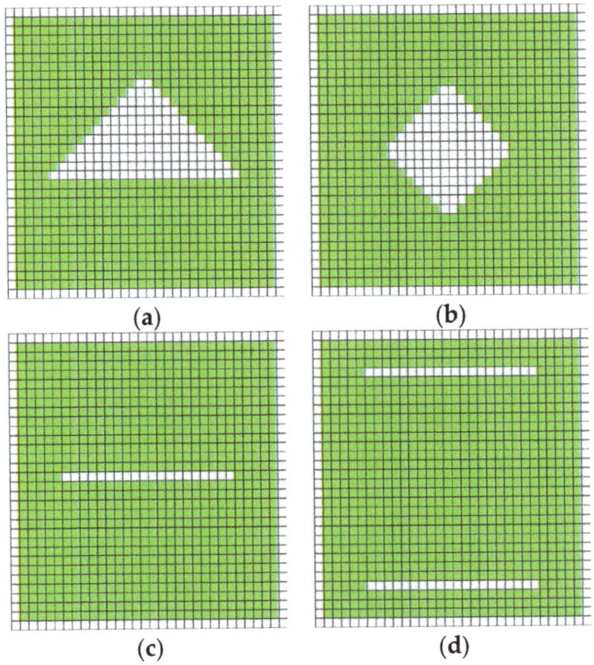

Figure 5. Testing the geometries of normalized multi-band patches: (**a**) triangle-slot patch, (**b**) circular slot patch, (**c**) single notch patch, and (**d**) double notch patch. Pixel dimensions: 5 mm × 5 mm. Green: conductive surface; white: uncovered substrate.

Since the feed-forward back-propagation network showed the best performance, we used this classifier in further tests.

3. Design Example

In order to demonstrate the functionality of the developed methodology, we designed a three-band antenna covering wireless local area network (WLAN) channels:

- 802.11b/g/n/ax: f_1 = 2.4 GHz;
- 802.11y: f_2 = 3.6 GHz;
- 802.11j: f_3 = 4.9 GHz.

The antenna should be designed for the substrate ARLON 25N with ε_r = 3.38, tan δ = 0.0025, and a height h = 1.524 mm. Since the whole patch is assumed to be in the half-wavelength resonance at f_1 = 2.40 GHz, we can evaluate the wavelength in the dielectrics according to [27]:

$$\lambda_d = \frac{c}{f_r\sqrt{\varepsilon_r}} \tag{2}$$

where f_r = 2.4 GHz, c is the velocity of light, and the dielectric constant equals to ε_r = 3.38. Numerically, λ_d = 68 mm, and the scaling factor related to the neural model equals to $n = \lambda_d/\lambda_0 = 68/300 = 0.227$.

Introducing the triplet of normalized frequencies [2.4/2.4; 3.6/2.4; and 4.9/2.4] to the input of the neural classifier, a G-slot antenna (Figure 2) is recommended as an optimum structure. The size of the patch W × L was 150 mm × 150 mm in the neural model and 34.1 mm × 34.1 mm in the recomputed model.

The coordinates of the vertices A through L of the polygonal slot (Figure 2) produced by the neural classifier are given in Table 1. The dimensions were recomputed using the scale n = 0.227 (Table 1), and a numerical model in HFSS was developed (Figure 6).

Table 1. Coordinates of vertices in the polygonal G slot in the patch depicted in Figure 2. Comparison of the neural model, the recomputed one, and the optimized one.

	Neural Network		Recomputed		Optimized	
A	−70.0	25.0	−15.9	5.7	−13.4	4.7
B	−60.0	25.0	−13.6	5.7	−11.6	4.7
C	−60.0	65.0	−13.6	14.8	−11.6	13.3
D	0.0	65.0	0.0	14.8	0.5	13.3
E	0.0	0.0	0.0	0.0	0.5	−0.5
F	−20.0	0.0	−4.5	0.0	−4.0	−0.5
G	−20.0	5.0	−4.5	1.1	−4.0	0.6
H	−5.0	5.0	−1.1	1.1	−0.6	0.6
I	−5.0	60.0	−1.1	13.6	−0.6	12.1
J	−55.0	60.0	−12.5	13.6	−10.5	12.1
K	−55.0	20.0	−12.5	4.5	−10.5	3.5
L	−70.0	20.0	−15.9	4.5	−13.4	3.5

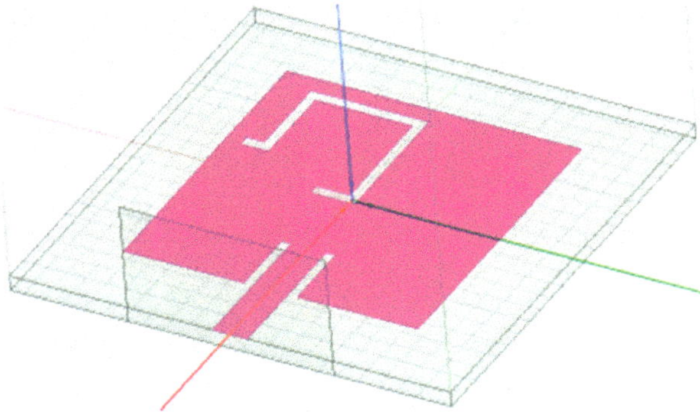

Figure 6. Numerical model of the G-slot antenna in HFSS. The patch is completed by a microstrip feeder and the wave port.

In the numerical model, the patch was completed by a microstrip feeder and a wave port (Figure 6). The antenna was simulated in the frequency range of 1.5 GHz–5.0 GHz. The impedance characteristics (Figure 7a) showed that resonance frequencies are shifted.

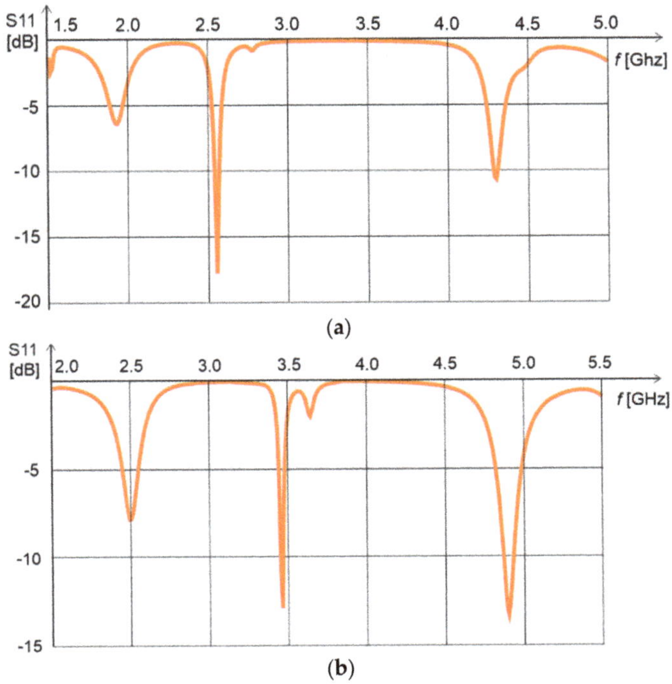

Figure 7. Impedance characteristics of the G-slot antenna: (**a**) initial model, (**b**) optimized model.

In a consequent step, Newton optimization was run to shift resonances towards the requested bands. After the optimization, patch dimensions were 30 mm × 30 mm. The vertices of the optimized slot polygon are given in Table 1. The impedance characteristics of the optimized antenna are depicted in Figure 7b. Obviously:

- f_1 = 2.4 GHz is shifted to 2.5 GHz and is not sufficiently deep.
- f_2 = 3.6 GHz corresponds to a shallow minimum, while the deep one is at 3.5 GHz.
- f_3 = 4.9 GHz is tuned successfully with $|S11| < -10$ dB.

4. Results

In this communication, we trained artificial neural networks on the approximate results of modal analysis. Neural models mapped a triplet of the lowest resonant frequencies to the corresponding layout of a patch. In order to develop sufficiently general neural models, antena structures were normalized. In total, 60 training patterns were created using 9 antenna layouts.

In MATLAB's Deep Learning Toolbox, we created a feed-forward ANN, a cascaded-forward ANN, and a probabilistic ANN. On four antenna layouts, which were not included in the training sets, the functionality of neural classifiers was tested. Whereas the probabilistic ANN failed, the feed-forward ANN showed relatively good results.

Using the feed-forward ANN, we tried to design a triple-band antenna covering WLAN bands 2.4 GHz, 3.6 GHz, and 4.9 GHz. The neural model returned a G-slot antenna as an optimum structure. The normalized antenna was recomputed for the substrate ARLON 25N, and a corresponding HFSS model was developed. Due to the shift of resonant frequencies with respect to the requested ones, the antenna was optimized by the Newton algorithm.

The optimized impedance characteristics are close to the requested ones, but the match is not perfect. Obviously, the concept of multiband antenna predesign by neural classifiers can work, but much more effort has to be devoted to the composition of a larger and more general training set comprising further families of antenna layouts. If sufficiently large training sets are composed, advanced deep-learning techniques can be applied. This is an intention for future work.

Author Contributions: Conceptualization Z.R. and J.V.; methodology Z.R. and J.V.; validation M.A.L.; investigation, Z.R., M.A.L. and J.V.; writing—original draft preparation Z.R.; writing—review and editing J.O.; visualization Z.R.; supervision Z.R. and J.V.; project administration Z.R. and J.O.; funding acquisition J.O. All authors have read and agreed to the published version of the manuscript.

Funding: The presented research was supported by the Czech Ministry of Defense (AIROPS, the University of Defense development program) and by the Internal Grant Agency of Brno University of Technology (FEKT-S-20-6526).

Conflicts of Interest: The authors declare no conflict of interest.

Abbreviations

ANN	Artificial Neural Network
CPU	Computer Processing Unit
CPW	Coplanar Waveguide
HFSS	High Frequency Field Simulator
PDE	Partial Differential Equation
WLAN	Wireless Local Area Network

References

1. Balanis, C.A. *Antenna Theory: Analysis and Design*, 4th ed.; John Wiley & Sons: Hoboken, NJ, USA, 2016; ISBN 978-1-1186-4206-1.
2. Haykin, S. *Neural Networks and Learning Machines: A Comprehensive Foundation*, 3rd ed.; Pearson Education: Upper Saddle River, NJ, USA, 2009; ISBN 978-0-1314-7139-9.
3. Elshamy, M.M.; Tiraturyan, A.N.; Uglova, E.V.; Elgendy, M.Z. Comparison of feed-forward, cascade-forward, and Elman algorithms models for determination of the elastic modulus of pavement layers. In Proceedings of the 4th International Conference on Geoinformatics and Data Analysis, Marseille, France, 14–16 April 2021. [CrossRef]
4. Raida, Z. Physical layer authentication of off-body channels by probabilistic neural networks. *Int. J. Numer. Model. Electron. Netw. Devices Fields* **2019**, *32*, e2628. [CrossRef]
5. Zhang, B.; Jin, C.; Cao, K.; Lv, Q.; Mittra, R. Cognitive conformal antenna array exploiting deep reinforcement learning method. *IEEE Trans. Antennas Propag.* **2022**, *70*, 5094–5104. [CrossRef]
6. Zhang, B.; Jin, C.; Cao, K.; Lv, Q.; Zhang, P.; Li, Y.; Li, M. Ultra-wide-scanning conformal heterogeneous phased array antenna based on deep deterministic policy gradient algorithm. *IEEE Trans. Antennas Propag.* **2022**, *70*, 5066–5077. [CrossRef]
7. Raida, Z. Modeling EM structures in the neural network toolbox of MATLAB. *IEEE Antennas Propag. Mag.* **2002**, *44*, 46–67. [CrossRef]
8. Pal, D.; Singhal, R.; Bandyopadhyay, A.K. Parametric optimization of complementary split-ring resonator dimensions for planar antenna size miniaturization. *Wirel. Pers. Commun.* **2022**, *123*, 1897–1911. [CrossRef]
9. Koziel, S.; Çalık, N.; Mahouti, P.; Belen, M.A. Accurate modeling of antenna structures by means of domain confinement and pyramidal deep neural networks. *IEEE Trans. Antennas Propag.* **2022**, *70*, 2174–2188. [CrossRef]
10. Zhou, Y.; Xie, J.; Ren, Q.; Zhang, H.H.; Liu, Q.H. Fast multi-physics simulation of microwave filters via deep hybrid neural network. *IEEE Trans. Antennas Propag.* **2022**, *70*, 5165–5178. [CrossRef]
11. Faustino, E.; Melo, M.C.; Buarque, P.; Bastos-Filho, C.J.A.; Arismar Cerqueira, S.; Barboza, E.A. Comparison of machine learning algorithms for application in antenna design. In Proceedings of the SBMO/IEEE MTT-S International Microwave and Optoelectronics Conference, Fortaleza, Brazil, 24–27 October 2021. [CrossRef]
12. Karahan, E.A.; Gupta, A.; Khankhoje, U.K.; Sengupta, K. Deep learning based modeling and inverse design for arbitrary planar antenna structures at RF and millimeter-wave. In Proceedings of the 2022 IEEE International Symposium on Antennas and Propagation and USNC-URSI Radio Science Meeting, Denver, CO, USA, 10–15 July 2022. [CrossRef]
13. Mir, F.; Kouhalvandi, L.; Matekovits, L. Deep neural learning based optimization for automated high performance antenna designs. *Sci. Rep.* **2022**, *12*, 16801. [CrossRef] [PubMed]
14. Nan, J.; Xie, H.; Gao, M.; Song, Y.; Yang, W. Design of UWB antenna based on improved deep belief network and extreme learning machine surrogate models. *IEEE Access* **2021**, *9*, 126541–126549. [CrossRef]
15. Johnson, J.M.; Rahmat-Samii, Y. Genetic algorithms in engineering electromagnetics. *IEEE Antennas Propag. Mag.* **1997**, *39*, 7–25. [CrossRef]
16. Weile, D.S.; Michielssen, E. Genetic algorithm optimization applied to electromagnetics: A review. *IEEE Trans. Antennas Propag.* **1997**, *45*, 343–353. [CrossRef]

17. Robinson, J.; Rahmat-Samii, Y. Particle swarm optimization in electromagnetics. *IEEE Trans. Antennas Propag.* **2004**, *52*, 397–407. [CrossRef]
18. Lu, J.; Ireland, D.; Lewis, A. Multi-objective optimization in high frequency electromagnetics—An effective technique for smart mobile terminal antenna (SMTA) design. *IEEE Trans. Magn.* **2009**, *45*, 1072–1075. [CrossRef]
19. Shi, D.; Lian, C.; Cui, K.; Chen, Y.; Liu, X. An intelligent antenna synthesis method based on machine learning. *IEEE Trans. Antennas Propag.* **2022**, *70*, 4965–4976. [CrossRef]
20. Vesely, J.; Olivova, J.; Gotthans, J.; Gotthans, T.; Raida, Z. Classification of microwave planar filters by deep learning. *Radioengineering* **2022**, *31*, 69–76. [CrossRef]
21. Cap, A.; Raida, Z.; Heras-Palmero, E.; Lamadrid-Ruiz, R. Multi-band planar antennas: A comparative study. *Radioengineering* **2005**, *14*, 11–20.
22. Chi, Y.J.; Chen, F.C. On-body adhesive-bandage-like antenna for wireless medical telemetry service. *IEEE Trans. Antennas Propag.* **2014**, *62*, 2472–2480. [CrossRef]
23. Ali, T.; Bhudevi, H.; Subhash, B.K.; Prasad, K.D.; Biradar, R.C. A miniaturized dual band antenna loaded with L and G-shaped slots for WiMAX/WLAN applications. In Proceedings of the 3rd IEEE International Conference on Recent Trends in Electronics, Information & Communication Technology (RTEICT), Bangalore, India, 18–19 May 2018. [CrossRef]
24. Hamd, H.I.; Mohamed, W.Q.; Ahmed, H.B. Design and simulation of H shape and duplicate U shape slots microstrip patch antenna for WiMAX applications. In Proceedings of the 5th International Symposium on Multidisciplinary Studies and Innovative Technologies (ISMSIT), Ankara, Turkey, 21–23 October 2021. [CrossRef]
25. Lacik, J.; Mikulasek, T.; Raida, Z.; Urbanec, T. Substrate integrated waveguide monopolar ring-slot antenna. *Microw. Opt. Technol. Lett.* **2014**, *56*, 1865–1869. [CrossRef]
26. Partial Differential Equation Toolbox. Available online: https://uk.mathworks.com/products/pde.html (accessed on 28 November 2022).
27. Garg, R.; Bhartia, P.; Bahl, I.; Ittipiboon, A. *Microstrip Antenna Design Handbook*; Artech House: Norwood, MA, USA, 2001; ISBN 978-0-8900-6513-6.
28. Deep Learning Toolbox. Available online: https://uk.mathworks.com/help/deeplearning (accessed on 28 November 2022).

electronics

MDPI

Article

Optimization and Design of Balanced BPF Based on Mixed Electric and Magnetic Couplings

Qiwei Li [1,2], Jinyong Fang [2,*], Wen Cao [3], Jing Sun [2], Jun Ding [1,*], Weihao Tie [2], Feng Wei [4], Chang Zhai [2] and Jiangniu Wu [2]

1 School of Electronics and Information, Northwestern Polytechnical University, Xi'an 710129, China
2 China Academy of Space Technology (Xi'an), Xi'an 710100, China
3 School of Electronics and Control Engineering, Chang'an University, Xi'an 710064, China
4 National Key Laboratory of Antennas and Microwave Technology, Xidian University, Xi'an 710071, China
* Correspondence: fangjy111@163.com (J.F.); dingjun@nwpu.edu.cn (J.D.)

Abstract: A balanced bandpass filter (BPF) with an improved frequency selectivity for differential-mode (DM) excitation and high rejection for common-mode (CM) excitation is proposed in this paper. Two half-wavelength stepped impedance resonators (SIRs) are employed based on mixed electric and magnetic couplings to realize a DM passband centered at 2.48 GHz. The center frequency and bandwidth can be easily controlled by optimizing the dimensions of SIRs and the coupling between them, respectively. Meanwhile, two transmission zeros (TZs) are generated based on the mixed electric and magnetic couplings and are independently controlled by tuning the coupling strength. Moreover, a wide DM stopband can be realized by optimizing the SIRs. The proposed balanced BPF is fed by balanced U-type microstrip–slotline transition structures, which can achieve high wideband CM rejection without influencing the DM responses, and the design complexity can be clearly reduced. Finally, a balanced BPF is fabricated, and a good agreement between the simulation and the measurement is observed, which verifies the design method.

Keywords: balanced; microstrip–slotline transition; mixed electric and magnetic couplings

Citation: Li, Q.; Fang, J.; Cao, W.; Sun, J.; Ding, J.; Tie, W.; Wei, F.; Zhai, C.; Wu, J. Optimization and Design of Balanced BPF Based on Mixed Electric and Magnetic Couplings. *Electronics* **2023**, *12*, 2125. https://doi.org/10.3390/electronics12092125

Academic Editor: Walter Ciccognani

Received: 28 February 2023
Revised: 23 April 2023
Accepted: 5 May 2023
Published: 6 May 2023

1. Introduction

With the rapid development of modern wireless technologies, multiservice communication systems, such as 4G and 5G, have become widespread, and these systems will begin to interfere with each other. Recently, balanced circuits are increasingly coming into focus due to their better performance, which includes good noise immunity, low electromagnetic interference and easy connection with other balanced components or antennas compared to single-end circuits [1,2]. Balanced bandpass filters (BPFs) can achieve high common-mode (CM) rejection according to the relevant research. Several balanced BPFs with improved CM suppression were designed in [3–7]. In [3], a balanced BPF was designed by using a center-loaded half-wavelength resonator in 2010. However, its differential-mode (DM) passband selectivity and CM rejection could be improved further. In [4], a low-loss wideband differential BPF was presented by using a pair of multilayer differential microstrip–slotline transitions and microstrip line resonators. However, the proposed BPF is difficult to fabricate and expensive. A balanced filter with a broadband DM transmission and a good CM block was fabricated based on a slotline resonator in [5], but the selectivity of the DM passband could be improved further. Recently, a compact second-order balanced BPF with high selectivity was presented based on magnetically coupled resonators in [6]. However, the DM stopband is narrow. In [7], an ultra-band BPF DM passband was achieved by employing an H-type slotline structure. In order to improve selectivity, a source-load coupling was introduced. Meanwhile, one more layer of substrate was employed to reduce the DM passband insert loss. Unfortunately, the CM suppression could be improved further. In [8,9], a balanced BPF was achieved based on right-angled isosceles triangular patch resonators and half-wavelength folded S-shaped slotline resonators, respectively. However, balanced BPFs with a compact size, good DM selectivity, low insertion loss and high CM suppression are still demanded.

A balanced BPF is designed with improved DM frequency selectivity and high CM rejection in this manuscript. The designed filter is mainly based on two half-wavelength stepped impedance resonators (SIRs), in which the SIRs are coupled by both an electric field and a magnetic field, respectively. The centered DM passband is formed at 2.48 GHz, which can be controlled by the dimensions of the SIRs. Meanwhile, two transmission zeros (TZs) are realized to achieve steep DM passband selectivity. TZs can be independently controlled by changing the coupling strength between the two SIRs. In addition, the CM suppression is broad and independent of the DM responses, which can clearly simplify the design procedure. Moreover, a broad DM stopband can also be achieved. Finally, a balanced BPF is fabricated, and a good agreement between the simulation and the measurement is observed, which verifies the design method.

2. Analysis and Design

2.1. Configuration of Balanced BPF

The traditional method to improve filter selectivity is to introduce an additional transmission path, which can generate transmission zeros. However, the size of the designed filter will increase. In this paper, the electromagnetic coupling method is used to introduce an extra transmission path while keeping the filter size unchanged. The 3D and plane structure of the designed balanced BPF is given in Figure 1. The yellow block stands for the microstrip line on the top surface of the substrate, whereas the white block and the blue block represent the slotline and ground on the bottom surface, respectively. It can be observed that the proposed component is composed of two quarter-wavelength SIRs, U-type microstrip–slotline transition structures and L-type microstrip–slotline transition structures. Generally, the design of a balanced component can be divided into two parts: DM part design and CM part design.

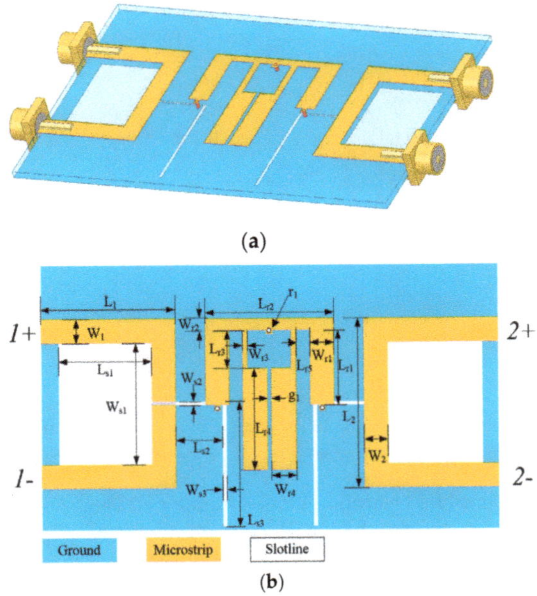

Figure 1. (**a**) 3D view and (**b**) top view of the proposed balanced BPF. (W_1 = 2.4, W_2 = 2.4, L_1 = 12.4, L_2 = 16, W_{s1} = 11.0, W_{s2} = 0.2, W_{s3} = 0.4, L_{s1} = 8.0, L_{s2} = 5.0, L_{s3} = 13.0, L_{r1} = 7.0, L_{r2} = 16.0, L_{r3} = 5.0, L_{r4} = 9.6, L_{r5} = 2.8, W_{r1} = 2.4, W_{r2} = 1.0, W_{r3} = 0.8, W_{r4} = 2.6, r_1 = 0.2 and g_1 = 0.4. All values are in millimeters.)

2.2. DM Response

A common SIR is composed of one high-impedance section and one low-impedance section, where Y_1, Y_2 and θ_1, θ_2 represent the characteristic admittance and electrical length of each stub, respectively, as shown in Figure 2a. The input admittance can be concluded as [10]

$$Y_{in} = Y_2 \frac{jY_1 \tan\theta_1 + jY_2 \tan\theta_2}{Y_2 + j(jY_1 \tan\theta_1)\tan\theta_2} \quad (1)$$

$$Y_{in} = Y_2 \frac{j(Y_1/Y_2)\tan\theta_1 + j\tan\theta_2}{1 - (Y_1/Y_2)\tan\theta_1 \tan\theta_2} \quad (2)$$

For convenience, the impedance ratio is defined as K, and the electrical length ratio is defined as α

$$K = \frac{Y_1}{Y_2} \quad (3)$$

$$\alpha = \frac{\theta_2}{\theta_1 + \theta_2} = \frac{\theta_2}{\theta_t} \quad (4)$$

According to the resonance condition $Y_{in} = 0$, (2) can be derived as:

$$\frac{K\tan\theta_1 + \tan\theta_2}{1 - K\tan\theta_1 \tan\theta_2} = 0 \quad (5)$$

Therefore, the resonance characteristic of SIR is determined by both impedance ratio K and length ratio α. Beyond that, by properly tuning K and α, high order harmonics can be adjusted. Figure 2b shows a regular change in the normalized ratios of the first spurious frequency (f_{s1}) to the fundamental resonance frequency (f_0) for SIR, for which the impedance ratio K increases from 0.1 to 1 when the length ratio α varies from 0 to 1. By adjusting impedance ratio K and length ratio α, the harmonics of SIR are far from the fundamental frequency to achieve a wide stopband feature. The initial simulation conditions are Z_1 = 50 Ohm, Z_2 = 20 Ohm, $\theta_1 = 90°$ and $\theta_2 = 60°$.

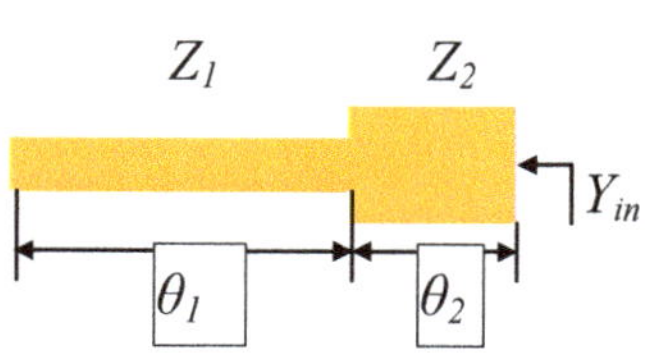

(a)

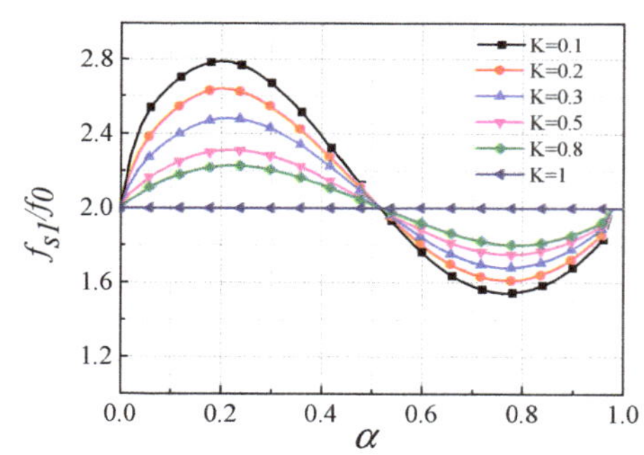

(b)

Figure 2. (**a**) Configuration and (**b**) normalized ratios of the first spurious frequency to the fundamental resonant frequency of SIR.

The simplified coupling topology under DM signal excitation is shown in Figure 3, in which the blue circles, S/S′ and L/L′, represent the microstrip transmission lines with a characteristic impedance of 50 Ohm; the white circles, N and N′, denote the slotline; and the yellow circles, l and l′, represent SIRs. The coupling between the microstrip transmission line and slotline is magnetic field coupling. However, there are two kinds of couplings between the two SIRs. The first kind of coupling is electrical field coupling, which is achieved by the parallel coupling between the low-impedance sections of SIRs. The second kind of coupling is magnetic field coupling, which is realized by a metallized vie hole between the high-impedance parts of SIRs. Figure 4 shows the electrical field and magnetic field distributions of the two SIRs at 2.47 GHz, which are achieved in EM simulation software HFSS. It can be seen that the electrical field coupling between the low-impedance parts of SIRs is strong, while the electrical field coupling between the high-impedance parts is weak at 2.47 GHz. Meanwhile, the magnetic field coupling between the high-impedance sections of SIRs is strong, while the magnetic field coupling between the low-impedance parts is weak at 2.47 GHz, which validates the aforementioned analysis. The electrical field coupling can be controlled by the gap between the low-impedance parts of SIRs, while the magnetic field coupling can be controlled by the diameter of the metallized via hole. The coupling between the slotline and microstrip line occurs by means of the magnetic field. The slotline, which is etched on one side of the substrate, is crossed at a right angle by a microstrip conductor on the opposite side. The microstrip and slotline extend about one quarter of a wavelength beyond the crossing point. To reduce the size, the extended quarter-wavelength open-circuit branch can also be replaced by a short-circuit point.

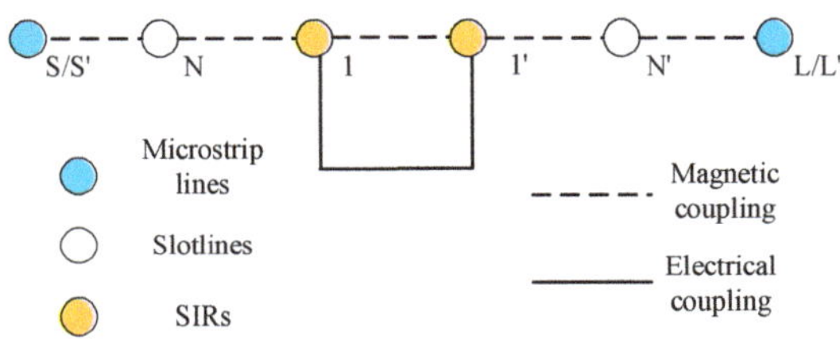

Figure 3. Simplified coupling topology under DM signal excitation of the proposed filter.

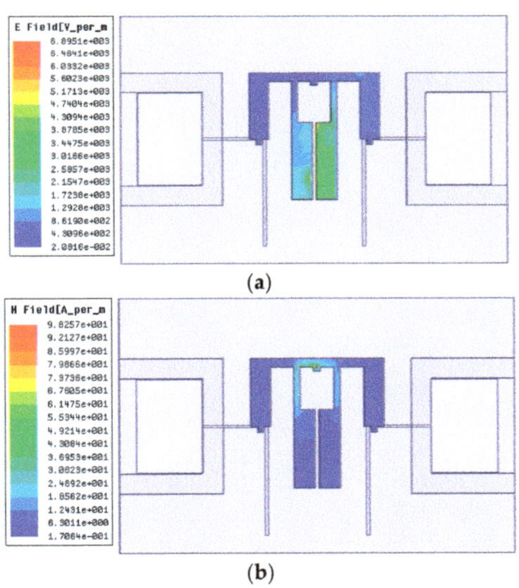

Figure 4. (**a**) Electrical field and (**b**) magnetic field distributions of the two SIRs at 2.47 GHz.

The performances of the DM and CM frequency characteristics of the proposed balanced filter with different dimensions are also analyzed by HFSS 13.0. As observed in Figure 5, the center frequency, bandwidth and the positions of *TZ* can be controlled by adjusting the dimensions. As observed in Figure 5a,b, the center frequency of the proposed balanced BPF is decreased when L_{r4} is increased, and the bandwidth is decreased when g_1 is increased. For the coupling resonators, the electrical coupling and magnetic coupling can be achieved by the gap between the high-impedance sections of SIRs and the metallized via hole. Due to the cancellation effect of the electrical coupling and magnetic coupling, TZ_l (the first transmission zero) is realized in the lower passband. Due to the U-type microstrip–slotline transition structure, TZ_h (the second transmission zero) is generated in the upper passband. Therefore, the selectivity of the DM passband can be improved greatly. As shown in Figure 5c, the position of TZ_l is controlled by changing r_1. Meanwhile, TZ_h is moved to high frequency when L_{r5} increases, as observed in Figure 5d. In addition, step impedance slotline structure is used to achieve the miniaturization of the proposed balanced BPF. Additionally, the impact of slotlines' width and length on the transmission performance is shown in Figure 5e. Therefore, the DM performances of the proposed balanced filter can be independently controlled by the related parameters.

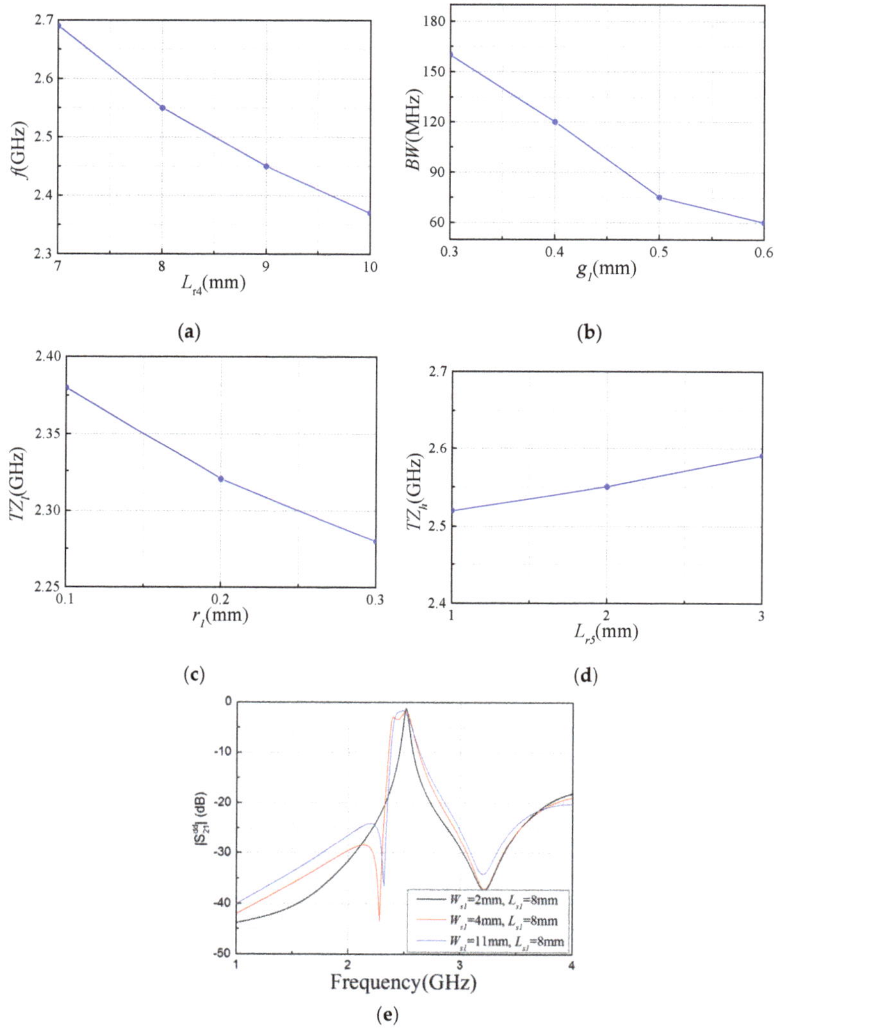

Figure 5. DM responses with different parameters: (**a**) L_{r4}; (**b**) g_1; (**c**) r_1; (**d**) L_{r5}; (**e**) W_{s1}. (W_1 = 2.4, W_2 = 2.4, L_1 = 12.4, L_2 = 16, W_{s1} = 11.0, W_{s2} = 0.2, W_{s3} = 0.4, L_{s1} = 8.0, L_{s2} = 5.0, L_{s3} = 13.0, L_{r1} = 7.0, L_{r2} = 16.0, L_{r3} = 5.0, L_{r4} = 9.6, L_{r5} = 2.8, W_{r1} = 2.4, W_{r2} = 1.0, W_{r3} = 0.8, W_{r4} = 2.6, r_1 = 0.2 and g_1 = 0.4. All values are in millimeters.)

2.3. CM Suppression

A slotline is a planar transmission structure first proposed by Cohn in 1968, which consists of a narrow slot etched in the metallization on one side of a dielectric substrate. The metal on the other side of the substrate is removed. This geometry is well suited for applications in microwave-integrated circuits. The slotline, which is etched on one side of the substrate, is crossed at a right angle by a microstrip conductor on the opposite side. The microstrip extends about one quarter of a wavelength beyond the slot. The transition can be fabricated using the usual photoetching process and is thus easily reproducible. Additionally, as the microstrip part of the circuit can be placed on one side of the substrate and the slotline part on the other side, this transition makes two-level circuit design possible. Coupling between the slotline and microstrip line occurs by means of the magnetic field. A balanced U-type microstrip/slotline transition, as depicted in Figure 6, comprises a U-type stepped-impedance microstrip feedline and a slotline resonator. The slotline resonator is etched on the bottom ground and the U-type microstrip line is on the top side of the substrate. The width and length of the slotline will influence the conversion efficiency from microstrip line to slotline, i.e., the insertion loss of the bandpass filter. Generally, the slotline extends about one quarter of a wavelength beyond the intersection of microstrip lines and slotlines. The electric field distributions excited by DM and CM signals in U-type microstrip–slotline transition structure are shown in Figure 6, in which the red arrow represents the electric field in microstrip line, the purple arrow represents the electric field in slotline and the symmetry plane is denoted by A-A′. When the balanced ports are excited by DM signals, the symmetric plane, A-A′, can be regarded as an electrical wall, as observed in Figure 6a. The DM signals along the U-type microstrip line can be converted easily into the slotline mode because of the strong magnetic coupling. However, when the circuit is under CM signal excitation, a virtual magnetic wall will be realized at the plane, A-A′, as observed in Figure 6b. Because the electric field of the slotline and the magnetic wall are vertical to each other, the CM signals cannot be coupled to the slotline. Moreover, the CM characteristic of the U-type microstrip–slotline transition structure is related to its own boundary condition and independent of the working frequency, as observed in Figure 7. Therefore, the design complexity can be reduced significantly. To reduce the size, the slotlines are designed as stepped impedance, and the specific dimensions are optimized by HFSS. The uniform-impedance slotline and the stepped-impedance slotline can both create good CM suppression. The length of the balanced microstrip–slotline transition structure can be effectively reduced by employing stepped-impedance slotline. Therefore, a stepped-impedance slotline structure is used to achieve the miniaturization of the proposed balanced BPF.

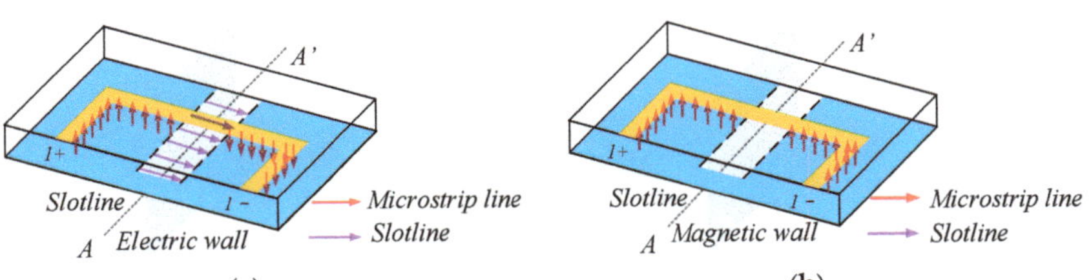

Figure 6. Electric field distributions of the balanced feed ports under (**a**) DM and (**b**) CM excitation.

2.4. Design Procedure

Based on the above analysis and discussions, a simple design procedure of the proposed balanced BPF can be summarized as follows:

Step (1): For the desired center frequency, the lengths of SIR can be selected appropriately according to Figure 5a.

Step (2): For the desired bandwidth, other dimensions of SIR can be selected appropriately according to Figure 5b–d.

Step (3): For the desired center frequency, the lengths of the slotlines can be chosen.

Step (4): All values are further optimized to realize better DM and CM responses of the proposed BPF. Full-wave electromagnetic simulation and dimension optimization in the commercial software of HFSS can be conducted.

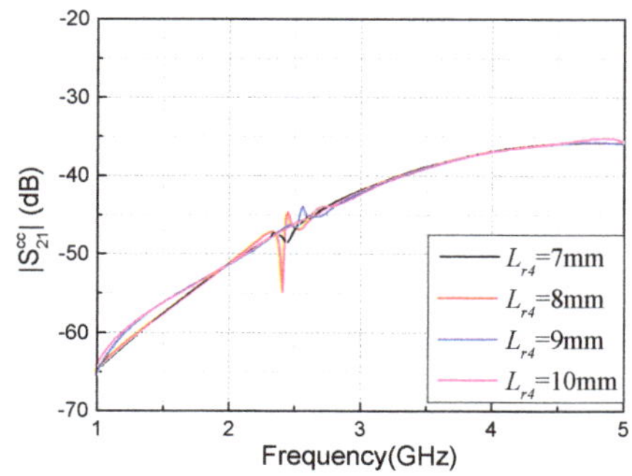

Figure 7. Simulated frequency responses with CM excitation of the proposed balanced filter for different L_{r4}. (W_1 = 2.4, W_2 = 2.4, L_1 = 12.4, L_2 = 16, W_{s1} = 11.0, W_{s2} = 0.2, W_{s3} = 0.4, L_{s1} = 8.0, L_{s2} = 5.0, L_{s3} = 13.0, L_{r1} = 7.0, L_{r2} = 16.0, L_{r3} = 5.0, L_{r5} = 2.8, W_{r1} = 2.4, W_{r2} = 1.0, W_{r3} = 0.8, W_{r4} = 2.6, r_1 = 0.2 and g_1 = 0.4. All values are in millimeters.)

3. Simulation and Measurement

The photographs of the fabricated balanced BPF are shown in Figure 8. The substrate used in this design is F4BM (ε_r =2.2 and h = 0.8 mm). Due to the machining accuracy, there is a certain deviation between the simulation and the measurement. The effective size is about 43.4 mm × 20.9 mm ($0.49\lambda_g \times 0.24\lambda_g$, where λ_g is the waveguide wavelength at 2.48 GHz). Figure 9 shows the simulation and the measurement. The measured S-parameters are carried out with an Agilent Network Analyzer N5230A. The measured DM center frequency is 2.48 GHz, with a 3-dB fractional bandwidth (FBW) of 4%. The measured minimum DM insertion loss for the center frequency is 1.74 dB, and the return loss is greater than 14 dB. Moreover, the DM stopband is extended to 10.8 GHz with a rejection level of 11.5 dB, which is 4.35 times the fundamental frequency, as shown in Figure 10. For the CM responses, a suppression of more than 40 dB can be achieved from DC to 6 GHz. Table 1 shows the comparison of the proposed balanced BPF with other proposed BPFs. It can be observed that the fabricated balanced BPF shows a better frequency selectivity of the DM passband and a higher CM rejection.

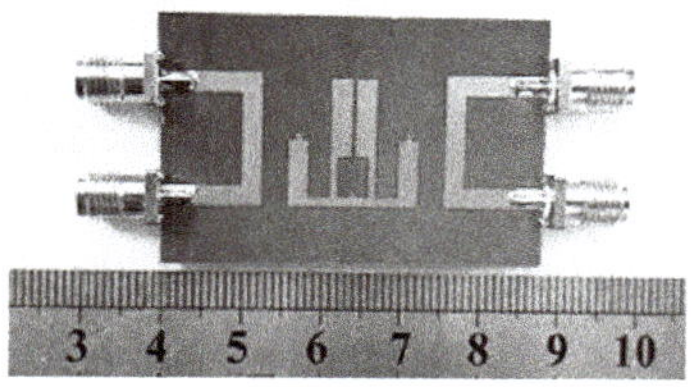

(**a**)

(**b**)

Figure 8. Photographs of the fabricated balanced BPF: (**a**) Top view; (**b**) bottom view.

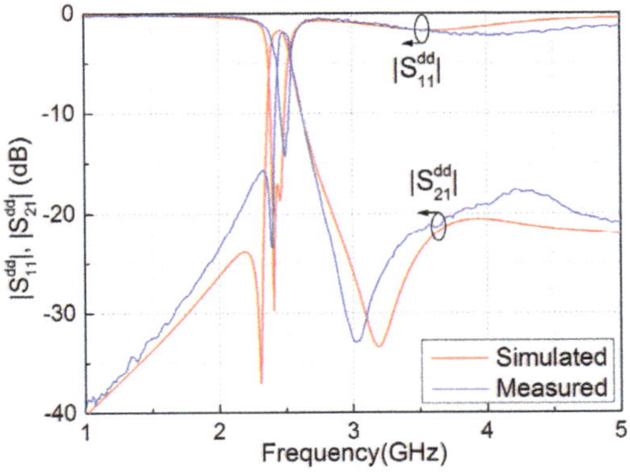

(**a**)

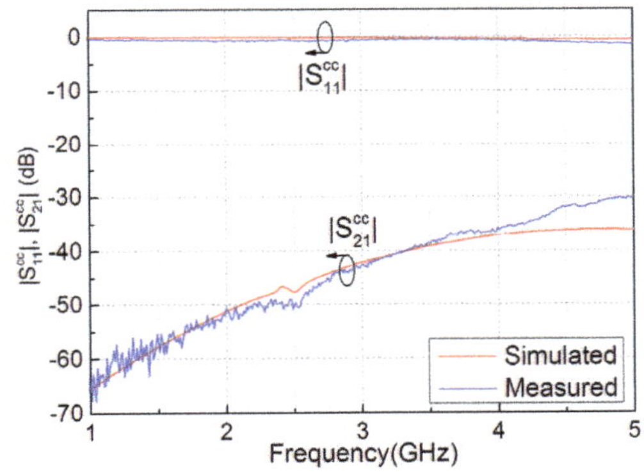

(**b**)

Figure 9. Comparisons between the simulated and measured results (narrowband): (**a**) DM responses; (**b**) CM responses. (W_1 = 2.4, W_2 = 2.4, L_1 = 12.4, L_2 = 16, W_{s1} = 11.0, W_{s2} = 0.2, W_{s3} = 0.4, L_{s1} = 8.0, L_{s2} = 5.0, L_{s3} = 13.0, L_{r1} = 7.0, L_{r2} = 16.0, L_{r3} = 5.0, L_{r4} = 9.6, L_{r5} = 2.8, W_{r1} = 2.4, W_{r2} = 1.0, W_{r3} = 0.8, W_{r4} = 2.6, r_1 = 0.2 and g_1 = 0.4. all values are in millimeters.)

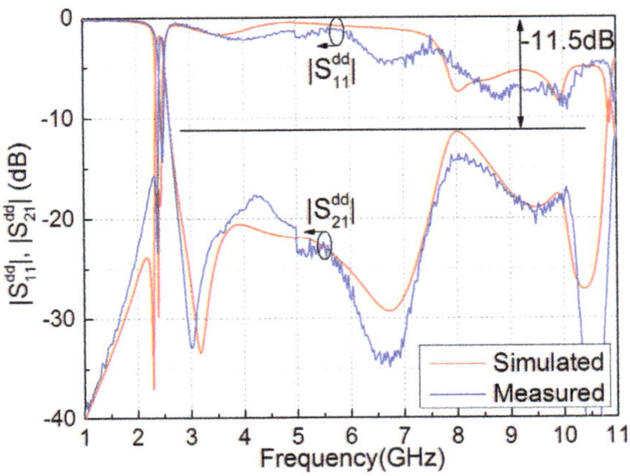

Figure 10. Comparisons between the simulated and measured results (wideband). (W_1 = 2.4, W_2 = 2.4, L_1 = 12.4, L_2 = 16, W_{s1} = 11.0, W_{s2} = 0.2, W_{s3} = 0.4, L_{s1} = 8.0, L_{s2} = 5.0, L_{s3} = 13.0, L_{r1} = 7.0, L_{r2} = 16.0, L_{r3} = 5.0, L_{r4} = 9.6, L_{r5} = 2.8, W_{r1} = 2.4, W_{r2} = 1.0, W_{r3} = 0.8, W_{r4} = 2.6, r_1 = 0.2 and g_1 = 0.4. All values are in millimeters.)

Table 1. Performance comparison with some published balanced BPFs.

Refs.	f_0(GHz)/3-dB FBW (%)	Number of TZs	CMS (dB) ※	Upper Stopband (GHz)	Location of TZs (GHz)
[5]	2.4/4	2	35	5.0	0/8.5
[8]	2/12.9	2	27.8	4.0	1.01/2.62
Ref. [9] Design I	1.2/20.1	1	55	2.0	0/1.75
Ref. [11] Case 2	6.85	82	0	40	None
This work	2.47/6.4	2	45	10.8	2.4/3.01

※ CMS = Common-mode suppression.

4. Conclusions

A balanced BPF based on a U-type microstrip–slotline transition structure and mixed electric and magnetic couplings is designed in this paper. The measured minimum DM insertion loss for the center frequency is 1.74 dB, and the return loss is greater than 14 dB. Two TZs located at 2.4 and 3.01 GHz are realized by introducing the electric coupling and the magnetic coupling into the design to achieve the steep DM passband selectivity. Moreover, the DM stopband is extended to 10.8 GHz with a rejection level of 11.5 dB, which is 4.35 times the fundamental frequency. The CM suppression level can be up to 40 dB from DC to 6 GHz. The designed balanced BPF exhibits good DM passband performance and wideband CM rejection, which will be valuable in fully balanced RF front ends.

Author Contributions: Q.L., J.F., W.C., J.S., W.T., J.D., C.Z. and J.W. designed and fabricated the proposed filter. Q.L. and F.W. contributed to the discussion and reviewed the manuscript. F.W. revised the manuscript. All authors have read and agreed to the published version of the manuscript.

Funding: This work was supported by the Foundation Enhancement Plan, the National Key Laboratory of Science and Technology on Space Microwave (No: 6142411332211), the National Natural Science Foundation of China (61803042) and the Fundamental Research Funds for the Central Universities, CHD (300102322103).

Data Availability Statement: The data presented in this study are available on request from the corresponding author.

Conflicts of Interest: The authors declare no conflict of interest.

References

1. Wei, F.; Yang, Z.-J.; Qin, P.-Y.; Guo, Y.J.; Li, B.; Shi, X.-W. A Balanced-to-Balanced In-Phase Filtering Power Divider with High Selectivity and Isolation. *IEEE Trans. Microw. Theory Tech.* **2018**, *67*, 683–694. [CrossRef]
2. Feng, W.; Che, W.; Xue, Q. New Balance-Applications for Dual-Mode Ring Resonators in Planar Balanced Circuits. *IEEE Microw. Mag.* **2019**, *28*, 15–23. [CrossRef]
3. Shi, J.; Xue, Q. Balanced Bandpass Filters Using Center-Loaded Half-Wavelength Resonators. *IEEE Trans. Microw. Theory Tech.* **2010**, *58*, 970–977. [CrossRef]
4. Shi, J.; Shao, C.; Chen, J.-X. Compact Low-Loss Wideband Differential Bandpass Filter With High Common-Mode Suppression. *IEEE Microw. Wirel. Compon. Lett.* **2013**, *23*, 480–482. [CrossRef]
5. Deng, H.; Zhao, Y.; He, Y.; Liu, H.; Wang, H. High selectivity and common-mode suppression wideband balanced bandpass filter using slotline resonator. *IET Microw. Antennas Propag.* **2015**, *9*, 508–516. [CrossRef]
6. Deng, H.-W.; Zhang, T.; Liu, F.; Xu, T. High Selectivity and CM Suppression Frequency-Dependent Coupling Balanced BPF. *IEEE Microw. Wirel. Compon. Lett.* **2018**, *28*, 413–415. [CrossRef]
7. Sun, J.; Li, G.R. A balanced ultra-wideband bandpass filter based on H-type sandwich slotline. *Int. J. RF Microw. Comput.-Aided Eng.* **2021**, *31*, e22611. [CrossRef]
8. Liu, Q.; Wang, J.; Zhang, G.; Zhu, L.; Wu, W. A New Design Approach for Balanced Bandpass Filters on Right-Angled Isosceles Triangular Patch Resonator. *IEEE Microw. Wirel. Compon. Lett.* **2019**, *29*, 5–7. [CrossRef]
9. Zou, X.T.; Yang, Z.J.; Wei, F.; Li, B.; Shi, X.W. Compact Balanced Single-Band and Dual-Band BPFs with Controllable Bandwidth Using Folded S-Shaped Slotline Resonators (FSSRs). *Frequenz* **2018**, *73*, 13–18. [CrossRef]

10. Zhang, X.Y.; Chen, J.-X.; Xue, Q.; Li, S.-M. Dual-Band Bandpass Filters Using Stub-Loaded Resonators. *IEEE Microw. Wirel. Compon. Lett.* **2007**, *17*, 583–585. [CrossRef]
11. Aliqab, K.; Hong, J. Wideband Differential-Mode Bandpass Filters with Stopband and Common-Mode Suppression. *IEEE Microw. Wirel. Compon. Lett.* **2020**, *30*, 233–236. [CrossRef]

electronics

MDPI

Article

A Balanced BPF with Wide Bandwidth and Steep Selectivity Based on Slotline Stub Loaded Resonators (SSLRs)

Yifei Liu [1], Wei Wu [1], Jinxi Li [1], Mo Zhao [1] and Feng Wei [2,*]

[1] National Key Laboratory of Intense Pulsed Radiation Simulation and Effect, Northwest Institute of Nuclear Technology, Xi'an 710024, China; liuyifei@nint.ac.cn (Y.L.); wuwei@nint.ac.cn (W.W.); lijinxi@nint.ac.cn (J.L.); zhaomo@nint.ac.cn (M.Z.)

[2] Xi'an Key Laboratory of Millimeter Wave and Terahertz Technology, Xidian University, Xi'an 710071, China

* Correspondence: fwei@mail.xidian.edu.cn

Abstract: A balanced bandpass filter (BPF) with a wide differential mode (DM) bandwidth and steep DM passband selectivity and high common-mode (CM) block based on slotline stub-loaded resonators (SSLRs) is designed in this brief. The proposed SSLR, which is composed of a half-wavelength slotline resonator and a shorted slot stub loaded at the symmetrical plane, is applied to achieve a broad DM passband with a center frequency at 5.1 GHz and a 3 dB bandwidth of 92%. Meanwhile, two transmission zeros (TZs) are introduced to realize a steep DM passband selectivity. The center frequency and bandwidth of the DM passband can be adjusted by changing the dimensions of SSLRs and the gaps between them. Meanwhile, one narrow notched band with a 3 dB bandwidth of 4.9% is realized by employing one coupled quarter-wavelength short-circuited stub at the input ports. In addition, the DM stopband with a rejection level of 10 dB can be extended to 20 GHz. The designed balanced broadband filter is stimulated by a U-shaped microstrip line to slotline transition structure, which can realize wideband a CM block with a high suppression level without influencing the DM one, and this can reduce the design complexity. In order to prove the idea, a balanced broadband filter is designed and measured. The predicted results of S parameters are compared with the measured ones, and a good agreement is achieved.

Keywords: bandpass filter (BPF); balanced; microstrip slotline structure; CM suppression; high selectivity; slotline stub-loaded resonators (SSLRs)

Citation: Liu, Y.; Wu, W.; Li, J.; Zhao, M.; Wei, F. A Balanced BPF with Wide Bandwidth and Steep Selectivity Based on Slotline Stub Loaded Resonators (SSLRs). *Electronics* **2023**, *12*, 3389. https://doi.org/10.3390/electronics12163389

Academic Editor: Djuradj Budimir

Received: 7 July 2023
Revised: 29 July 2023
Accepted: 5 August 2023
Published: 9 August 2023

1. Introduction

In recent years, balanced circuits have been paid increasingly close attention due to their high immunity to environmental noise, interference, and crosstalk compared with their single-ended counterparts. Bandpass filters (BPFs) are essential devices in the RF front end of wireless communication systems. Therefore, balanced BPFs have been research hotspots, and several balanced BPFs were presented in [1–3]. In [2], three half-wavelength transmission lines were employed to achieve a balanced BPF with broad bandwidth. Two balanced BPFs using open stubs were proposed in [3]. However, the common-mode (CM) rejection can only be greater than 20 dB when covering the entire differential mode (DM) passband. A dielectric resonator (DR) with two modes located below the metal cavity was studied to realize a balanced filter with a dual-band performance in [4]. In recent years, a U-shaped microstrip line to slotline transition structures were proposed, which show high and wideband CM rejection [4,5]. The DM signals can be transmitted from the microstrip line to the non-resonant slotline while CM signals are suppressed. In order to generate a broad DM passband, different methods have been attempted and studied [6–8]. Stub-loaded resonators (SLRs) were employed to implement a wide DM passband in [6]. Different complementary split-ring resonators with a half wavelength were also employed to generate more DM resonant points [7]. However, it is difficult to implement a wide DM passband because the coupling between the resonators is weak.

In [8], novel strip-loaded slotline resonators are used to design differential wideband BPFs with intrinsic CM suppression. Five transmission poles (TPs) are generated while the flexibility of transmission zeros (TZs) is not mentioned. However, the stopband is narrow. On the other hand, the extant unexpected narrow band jamming signal, such as a wireless local-area network signal, may interfere with the ultra-wide band (UWB) users. Thus, a balanced UWB filter with a notch-band is expected to block these jamming signals.

In this brief, a balanced BPF with a wide DM bandwidth is designed. The proposed BPF can realize steep DM passband selectivity, controllable DM notched band, wide DM stopband, and wideband CM suppression with a high suppression level. The balanced BPF consist of two U-shaped microstrip lines to slotline transition structures and two sets of slotline stub-loaded resonators (SSLRs). The SSLRs are used to achieve a wide DM passband with a center frequency of 5.1 GHz. Meanwhile, two transmission null points are introduced to achieve a high DM frequency selectivity. Meanwhile, one narrow notched band with a 3 dB bandwidth of 4.9% is achieved to suppress the strong interference signal. On the other hand, the DM stopband with a rejection level of 10 dB can be extended to 20 GHz. In addition, the CM suppression is broad and is independent from the DM response. Therefore, the design complexity is reduced greatly. The simulated results are compared with the measured ones, and a good agreement is observed.

2. Balanced Wideband BPF

2.1. Overall Architecture

The designed balanced wideband filter is observed in Figure 1, and is composed of SSLRs, two U-shaped microstrip line to slotline transition structures, and quarter-wavelength short-circuited resonators. The orange block is the microstrip line located on the top surface of the substrate, while the white one shows the slotline on the bottom surface. For a balanced wideband BPF, there are two main parts: the DM part and the CM part. The wide DM passband is implemented by employing the proposed SSLRs. Figure 2 shows the coupling scheme of the proposed balanced wideband BPF. In order to generate two TZs, the pseudo-interdigital coupling between the proposed SSLRs is employed, which can improve the passband selectivity. U-shaped microstrip lines to slotline transition structures are used to implement a good and wideband CM suppression. Meanwhile, the CM responses are independent from the DM ones, which can simplify the design procedure. Moreover, two quarter wavelength shorted circuit resonators are loaded adjacent to the input ports to generate the notched band. The proposed balanced BPF is realized based on the Rogers RT/Duroid 5880 [9]. (ε_r = 2.2, tanδ = 0.0009, h = 0.8 mm). The dimensions are shown in Table 1.

Table 1. Dimensions of the designed balanced wideband filter. Unit: mm.

W_1	2.4	W_{s5}	0.15	L_{s1}	8.0	G_1	0.1
W_2	2.0	W_{s6}	0.5	L_{s2}	2.0	G_2	0.2
W_3	4.0	W_{s7}	0.2	L_{s3}	1.5	G_3	0.3
W_{s1}	5.2	W_{s8}	0.15	L_{s4}	10.0	L_{r1}	7.1
W_{s2}	0.12	L_1	12	L_{s5}	10.0	L_{r2}	2.8
W_{s3}	0.65	L_2	11	L_{s6}	10.2	W_{r1}	0.5
W_{s4}	0.15	L_3	10	L_{s7}	5.0	W_{r2}	0.4
R_1	0.2	L_4	6.75				

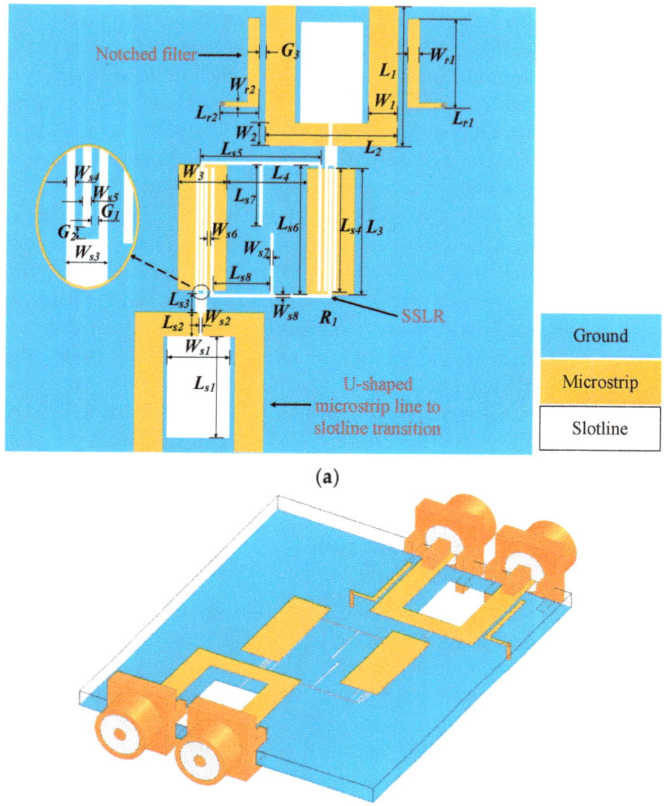

Figure 1. Layout of the designed balanced broadband filter: (**a**) Top view; (**b**) 3D view.

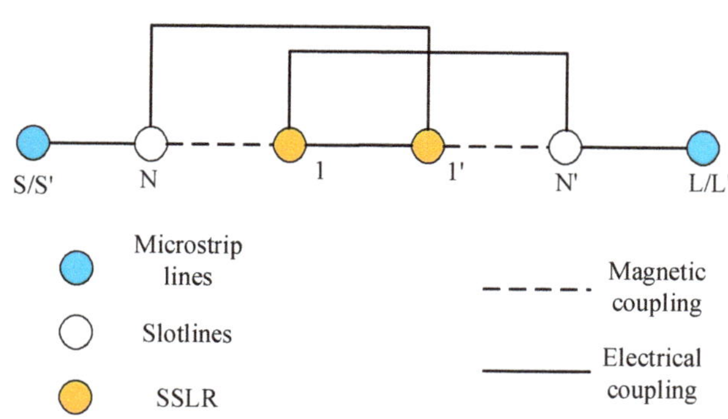

Figure 2. Coupling scheme of the designed balanced wideband BPF.

2.2. SSLR

Figure 3a shows the conventional SLR, which is composed of a microstrip half-wavelength resonator and a shorted stub located at the center. Similar to the structure of SLR, SSLR is composed of a half-wavelength resonator and a shorted slot stub, while it is located at ground level, as given in Figure 3b. Since the SLR is a symmetric structure, the odd-even mode method can be used to analyze it. The resonant frequencies of the even and odd modes of SLR can be calculated as follows:

$$f_e = \frac{c}{4(L_{r1}/2 + L_{r2})\sqrt{\varepsilon_{eff}}} \tag{1}$$

$$f_o = \frac{c}{8L_{r1}\sqrt{\varepsilon_{eff}}} \tag{2}$$

where c is the light speed in free space and ε_{eff} represents the effective dielectric constant of the substrate. Since SSLR is symmetrical in structure, an odd- and even-mode analysis method can also be employed. Similar to SLR, SSLR also has two resonant frequencies (f_e, f_o) that can be adjusted independently by controlling the dimensions. Figure 4 shows the resonant frequencies of the proposed SSLR with L_{s1} and L_{s2}, which are simulated by the software High Frequency Structural Simulator (HFSS). It can be seen from Figure 4 that the two resonant frequencies are changed to the lower frequency by increasing L_{s1}, while f_o can be independently controlled by L_{s2}. To further verify the validity of our analytical method, SSLR is analyzed through the characteristic mode analysis (CMA) with CST Microwave Studio [10]. The working principle of arbitrarily shaped structures can be intuitively provided without being affected by external excitation sources. During the simulation process, SSLR is etched on the bottom layer of the dielectric. The results show that two characteristic modes are generated from 2 to 10 GHz. The two characteristic modes, named modes 1 and 2, are implemented at 2.5 and 3.7 GHz, respectively. The electric field distribution of SSLR proposed in a different mode is shown in Table 2. The conclusion obtained by CST simulation is consistent with that obtained by HFSS simulation.

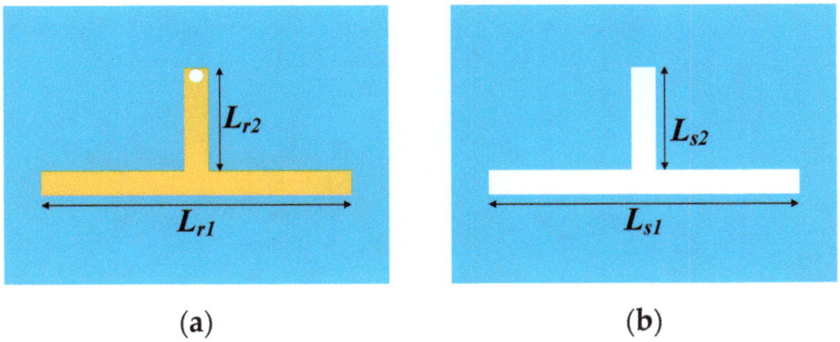

Figure 3. Configuration of (**a**) SLR and (**b**) SSLR.

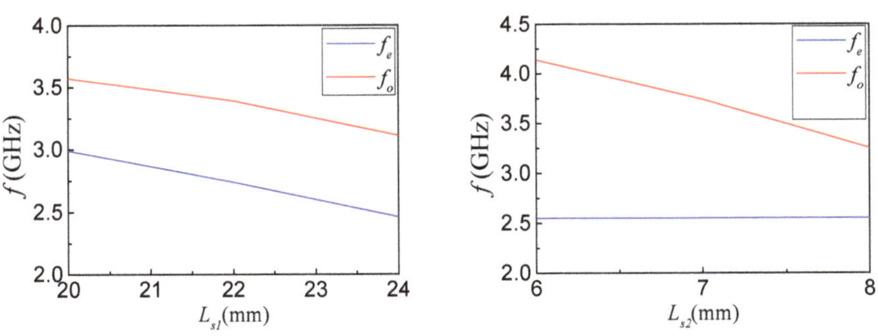

Figure 4. Resonant frequencies of the proposed SSLR with (**a**) L_{s1}, and (**b**) L_{s2}.

Table 2. E-field distribution of SSLR.

Mode	Mode 1 (2.5 GHz)	Mode 2 (3.7 GHz)
E-fields (V/m) 60 Max  0 Min	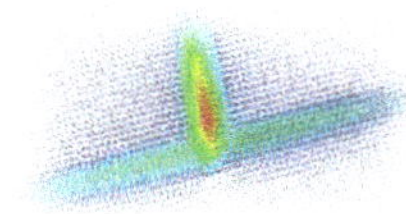	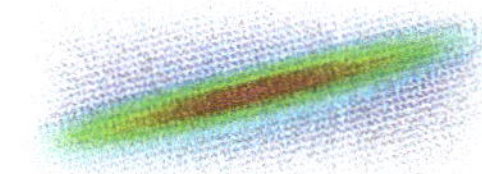

2.3. DM Transmission

In [10], the designed balanced BPF is realized by employing four microstrip lines to slotlines conversion, which results in the increased insertion loss and larger size. In this work, the proposed BPF is based on SSLRs, and two microstrip lines to slotline conversions are needed. It is easier to achieve lower loss and miniaturization. At the same time, a three-slotline couple structure is employed. Generally, the length of the slotline is a quarter wavelength. In order to increase the coupling strength between the three slotlines, a rectangular metal is used on the other side, as given in Figure 5a. The main parameters of the enhanced coupled structure are the characteristic impedance (Z_s) and the coupling capacitance (C_s), as given in Figure 5b. By introduing a back shielding strip, the per-unit-length shunt capacitance can be increased from C_s to (C_s + 2 × C_g) while Z_s will be decreased. Meanwhile, its transmission performance is shown in Figure 6. Through adjusting the dimension of the rectangular metal, different characteristic impedance (Z_s) and coupling capacitance (C_s) can be achieved. Therefore, the coupling strength of the three-slotline coupled structure can be changed.

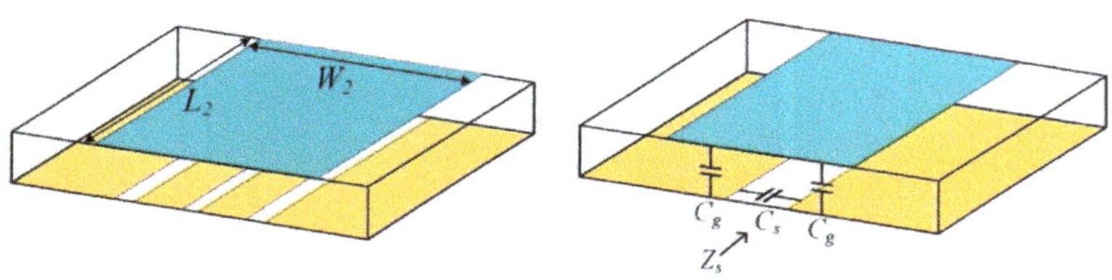

Figure 5. (**a**) Physical dimensions of the uniform slotline of the three-slotline coupled structure shielded by a rectangular metal. (**b**) Equivalent capacitance structure.

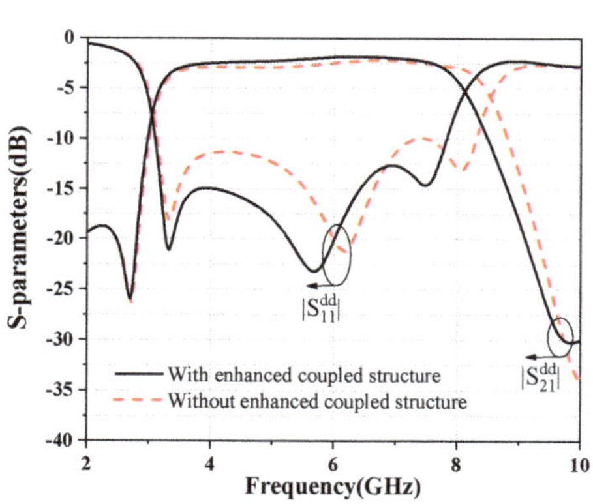

Figure 6. Comparison of S parameters with a different coupled structure.

2.4. DM Notched Band

The DM notched band is realized by introducing a folded quarter-wavelength resonator. The resonator is short-circuited and can be equivalent to one parallel series LC resonance circuit. The frequency responses of the coupled short-circuited quarter-wavelength resonator with different lengths are simulated by HFSS, as given in Figure 7. The simulated results show that the resonant frequency and the bandwidth can be controlled independently by adjusting the length L of the coupling resonator and the gap between the coupled resonator and the microstrip line. Therefore, by appropriately adjusting the dimensions of the short-circuited resonator, the notched band can be achieved at the desired frequency.

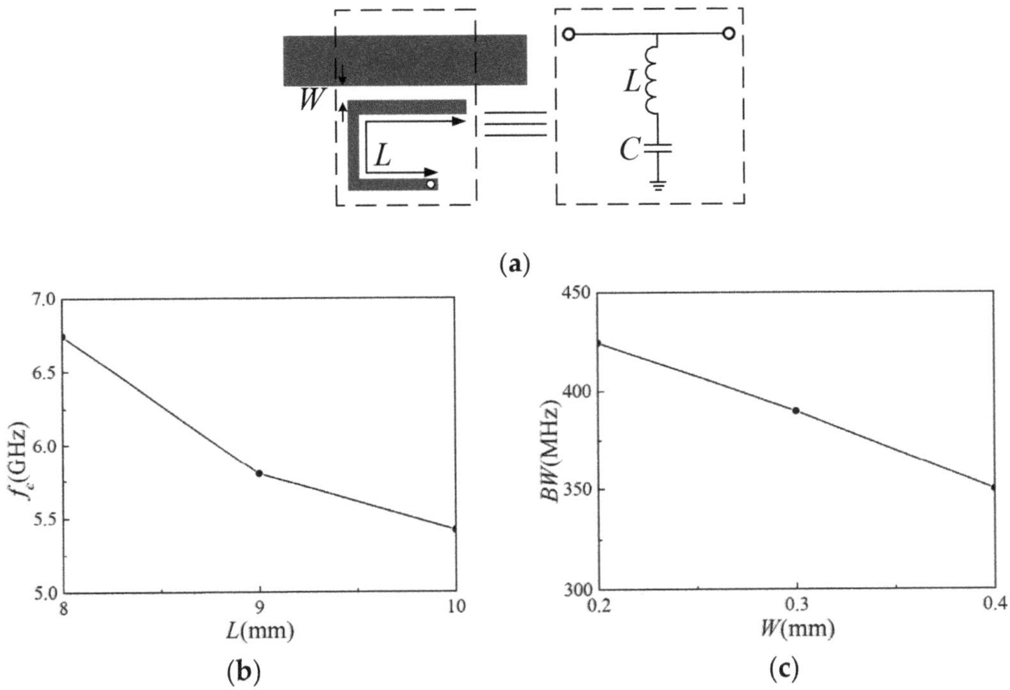

Figure 7. (**a**) Layout and equivalent circuit, (**b**) center frequency on L length, and (**c**) bandwidth of the coupled resonator with varied parameters.

2.5. CM Suppression

The electric field distributions in the U-shaped microstrip line to the slotline transition structure under DM and CM operation are shown in Figure 8, respectively, where A-A′ is the symmetry plane of the U-shaped microstrip line to slotline transition structure. When a DM excitation is applied, the plane of symmetric A-A′ can be equivalent to an electrical wall. The DM signals in the U-type microstrip line can be transmitted into the slotline modes with strong magnetic coupling, and then transformed to the next one. On the other hand, when a U-shaped microstrip line to slotline transition structure are under the CM operation, a virtual magnetic wall can be implemented at the symmetric plane A-A′. The magnetic wall is perpendicular to the electric field of the slotline mode, which conflicts with the magnetic wall's boundary condition. Therefore, the slotline cannot achieve the transmission of CM signals. So, the CM signals will be rejected. The suppression characteristic of the U-shaped microstrip line to slotline transition structure on CM signals is inherent and independent of the operating frequency, as shown in Figure 9. As a result, the design complexity can be reduced significantly.

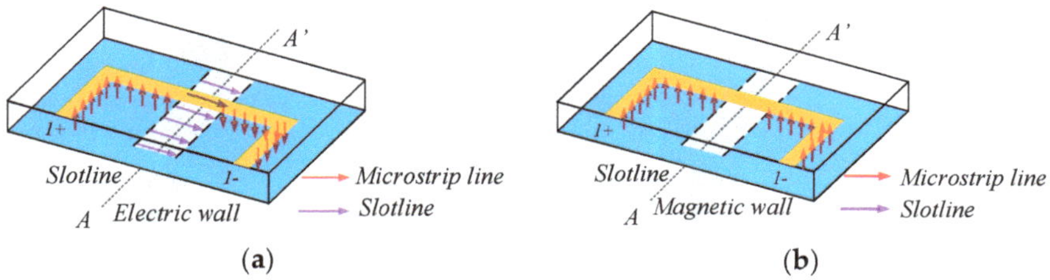

Figure 8. Electric field distributions in U-shaped microstrip line to slotline transition structure under (**a**) DM and (**b**) CM operation.

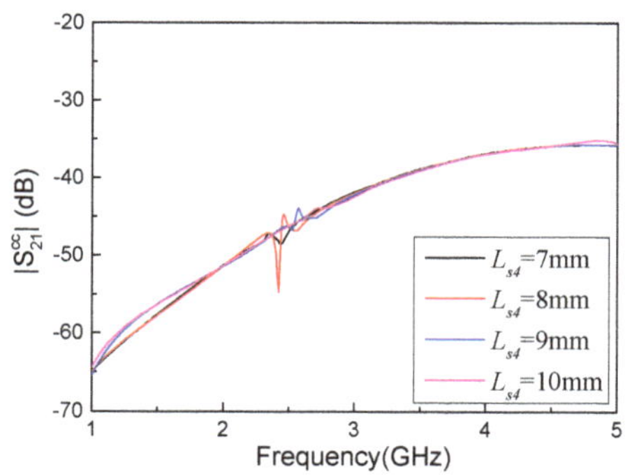

Figure 9. Simulated CM response of the proposed balanced BPF with L_{s4}.

2.6. Simulated and Measured S-Parameters

In order to prove the design idea, a balanced broadband filter is manufactured and measured, as given in Figure 10. The total effective dimension is only 33.2 mm × 25 mm. Figure 11 shows the comparison of the simulated results and the measured ones. The S-parameters of the proposed filter were measured with an Agilent vector network analyzer (VNA) N5230A. The measured DM passband center frequency is 5.1 GHz, with a 3-dB FBW of 92%. The insertion loss of the center frequency is better than 1.9 dB. Meanwhile, one narrow notched band with a 3 dB bandwidth of 4.9% is realized by employing one coupled quarter-wavelength short-circuited resonator at the input ports. In addition, the DM stopband can be extended to 20 GHz with a rejection level of 10 dB, as shown in Figure 12. For the CM suppression, the rejection level is more than 30 dB, covering the DM passband. The discrepancy between the measurement and the simulation results is due to the error of simulation as well as the error of PCB fabrication and measurement. We compared the performances of the proposed filter and the other ones, as given in Table 3. It can be observed that the designed balanced broadband filter shows better performances in the DM passband selectivity and CM block.

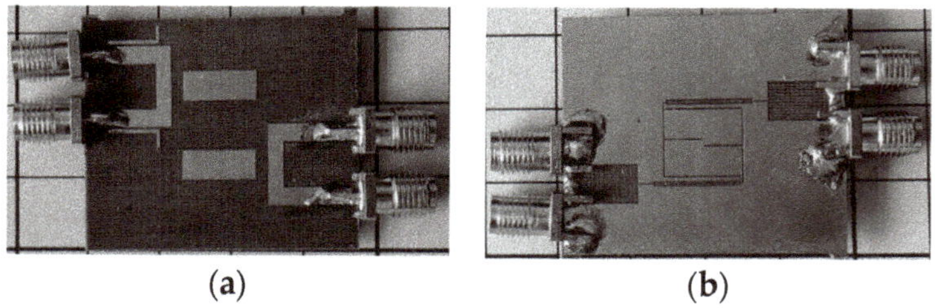

Figure 10. Fabricated balanced wideband filter: (**a**) Top; (**b**) Bottom.

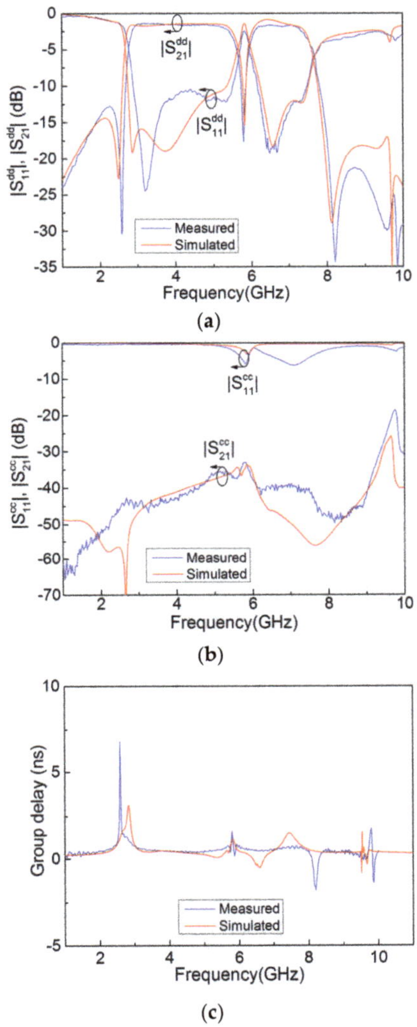

(a) (b) (c)

Figure 11. Comparison of the simulation and measurement (1–10 GHz): (**a**) DM reflection coefficient and transmission coefficient; (**b**) CM reflection coefficient and transmission coefficient; (**c**) DM group delay.

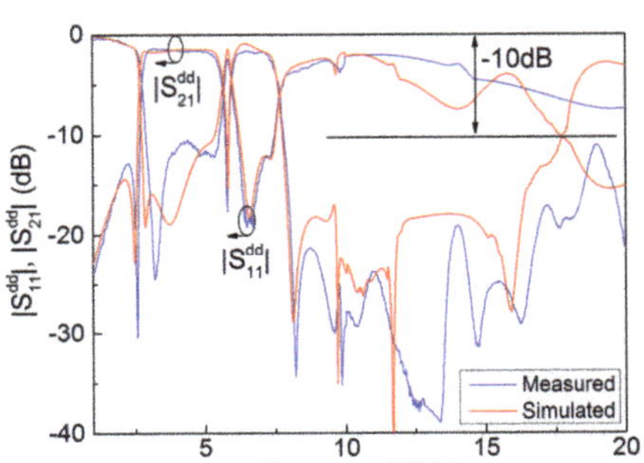

Figure 12. Comparisons of the simulated and measured DM reflection coefficient and transmission coefficient (1–20 GHz).

Table 3. Comparison with some reported balanced wideband BPFs.

Refs.	f_0 (GHz)/ 3-dB FBW (%)	$\Delta f_{3dB}/\Delta f_{20dB}$ *	CMS (dB)	Stopband (GHz)
[11]	2/55.5	0.53	25.38	5.0
[2]	1.8/57.8	0.47	20	4.5
[12]	1.3	0.58	45	7.0
[13] Case 2	6.85/58	0. 6	40	11
[14]	2/71.5	0.63	33	3.5
This work	5.1/92	0.73	34	20

* $\Delta f_{3dB}/\Delta f_{20dB}$ = 3-dB bandwidth/20-dB bandwidth. CMS = Common-mode suppression. FBW [15].

3. Conclusions

In this paper, an original balanced filter with broad bandwidth based on a U-shaped microstrip line to slotline transition structure and SSLR is achieved. By employing the pseudo-interdigital coupling structure in the feeding and SSLR, the better DM insertion losses of 1.9 dB, wider FBW of 92%, and steeper DM selectivity of 0.73 are realized. In addition, the DM stopband can be extended to 20 GHz with a rejection level of 10 dB. The U-shaped microstrip line to slotline transition structure can achieve wideband CM suppression with a high rejection level of more than 30 dB. The proposed balanced wideband BPF exhibits good DM transmission and CM rejection, which illustrates the potential implementation for future balanced systems.

Author Contributions: Y.L., W.W., J.L. and M.Z. designed and fabricated the proposed structure. F.W. revised the manuscript. All authors have read and agreed to the published version of the manuscript.

Funding: This research received no external funding.

Data Availability Statement: Not applicable.

Conflicts of Interest: The authors have no conflict of interest.

Abbreviations

BPF	Bandpass filter
CMA	Characteristic mode analysis
CM	Common-mode
DM	Differential mode
DR	Dielectric resonator
HFSS	High Frequency Structural Simulator
SSLRs	Slotline Stub Loaded Resonators
SLRs	Stub loaded resonators
TPs	Transmission poles
TZs	Transmission zeros
VNA	Vector network analyzer
UWB	Ultra-wide Band

References

1. Rich, W.F.; Stringer, T.A. System-Generated Electromagnetic Pulse and Spacecraft Charging Effects: A Review of the Technology as Applied to System Hardening Problems. *IEEE Trans. Nucl. Sci.* **1980**, *27*, 1523–1528. [CrossRef]
2. Qiang, J.; Shi, J.; Cao, Q.; Shen, Z. Compact Differential Wideband Bandpass Filters Based on Half-Wavelength Lines. *IEEE Microw. Wirel. Compon. Lett.* **2017**, *27*, 906–908. [CrossRef]
3. Feng, W.; Gao, X.; Che, W.; Yang, W.; Xue, Q. High Selectivity Wideband Balanced Filters with Multiple Transmission Zeros. *IEEE Trans. Circuits Syst. II Express Briefs* **2017**, *64*, 1182–1186. [CrossRef]
4. Chen, J.-X.; Li, J.; Shi, J. Miniaturized Dual-Band Differential Filter Using Dual-Mode Dielectric Resonator. *IEEE Microw. Wirel. Compon. Lett.* **2018**, *28*, 657–659. [CrossRef]

5. Xu, L.; Wang, Z.X.; Wei, F.; Li, R.; Zou, X.T. A Highly Selective Balanced Wideband Bandpass Filter Based on Nested Split-ring rResonators. *Int. J. RF Microw. Comput. -Aided Eng.* **2019**, *29*, e21875. [CrossRef]
6. Wei, F.; Yang, Z.-J.; Qin, P.-Y.; Guo, Y.J.; Li, B.; Shi, X.-W. A Balanced-to-Balanced In-Phase Filtering Power Divider with High Selectivity and Isolation. *IEEE Trans. Microw. Theory Tech.* **2019**, *67*, 683–694. [CrossRef]
7. Wei, F.; Guo, Y.J.; Qin, P.-Y.; Shi, X.W. Compact Balanced Dual- and Tri-band Bandpass Filters Based on Stub Loaded Resonators. *IEEE Microw. Wirel. Compon. Lett.* **2015**, *25*, 76–78. [CrossRef]
8. Guo, X.; Zhu, L.; Wu, W. Strip-Loaded Slotline Resonators for Differential Wideband Bandpass Filters with Intrinsic Common-Mode Rejection. *IEEE Trans. Microw. Theory Tech.* **2016**, *64*, 450–458. [CrossRef]
9. Available online: https://www.rogerscorp.cn/advanced-electronics-solutions/rt-duroid-laminates/rt-duroid-5880-laminates (accessed on 6 June 2023).
10. Available online: https://www.3ds.com/products-services/simulia/products/cst-studio-suite/ (accessed on 6 June 2023).
11. Zhang, W.; Wu, Y.; Liu, Y.; Yu, C.; Hasan, A.; Ghannouchi, F.M. Planar Wideband Differential-Mode Bandpass Filter with Common-Mode Noise Absorption. *IEEE Microw. Wirel. Compon. Lett.* **2017**, *27*, 458–460. [CrossRef]
12. Feng, L.-P.; Zhu, L. Strip-Loaded Slotline Resonator for Compact Differential-Mode Bandpass Filters with Improved Upper Stopband Performance. *IEEE Microw. Wirel. Compon. Lett.* **2017**, *27*, 108–110. [CrossRef]
13. Aliqab, K.; Hong, J. Wideband Differential-Mode Bandpass Filters with Stopband and Common-Mode Suppression. *IEEE Microw. Wirel. Compon. Lett.* **2020**, *30*, 233–236. [CrossRef]
14. Chen, W.; Wu, Y.; Wang, W. Planar Wideband High-Selectivity Impedance-Transforming Differential Band-pass Filter with Deep Common-Mode Suppression. *IEEE Trans. Circuits Syst. II Express Briefs* **2020**, *67*, 1914–1918.
15. Wang, X.-H.; Zhang, H.; Wang, B.-Z. A Novel Ultra-Wideband Differential Filter Based on Microstrip Line Structures. *IEEE Microw. Wirel. Compon. Lett.* **2013**, *23*, 128–130. [CrossRef]

MDPI
St. Alban-Anlage 66
4052 Basel
Switzerland
www.mdpi.com

Electronics Editorial Office
E-mail: electronics@mdpi.com
www.mdpi.com/journal/electronics

www.ingramcontent.com/pod-product-compliance
Lightning Source LLC
Chambersburg PA
CBHW061606210926
51850CB00009B/2226
9783725813094